D0824783

METHODS OF HYDROBIOLOGY

(FRESHWATER BIOLOGY)

METHODS OF HYDROBIOLOGY

(FRESHWATER BIOLOGY)

BY

JÜRGEN SCHWOERBEL

DRAWINGS, UNLESS OTHERWISE STATED, BY

WALTER SÖLLNER

PERGAMON PRESS

Oxford · London · Toronto

Sydney · Braunschweig

Pergamon Press Ltd., Headington Hill Hall, Oxford
Pergamon Press Inc., Maxwell House, Fairview Park, Elmsford, New York 10523
Pergamon of Canada Ltd., 207 Queen's Quay West, Toronto 1
Pergamon Press (Aust.) Pty. Ltd., 19a Boundary Street, Rushcutters Bay, N.S.W. 2011, Australia
Vieweg & Sohn GmbH, Burgplatz 1, Braunschweig

First English edition 1970
Reprinted 1972

The German original book was published by the FRANCK'SCHEN VERLAGSHANDLUNG, KOSMOS-VERLAG, Stuttgart, with the following original title '*Methoden der Hydrobiologie*'

Library of Congress Catalog Card No. 71-114573

REPRINTED BY A. WHEATON & CO., EXETER

08 006604 6

CONTENTS

PREFACE ix

SHORT INTRODUCTION TO THE HYDROBIOLOGY OF FRESH WATER 1

I. Types of Water 1

II. Levels of Correlation 1
1. Water as an Environment for Organisms 1
2. Community Life in Water 2
3. Total Life in Fresh Water 3

CHAPTER 1. SOME PHYSICAL AND CHEMICAL METHODS FOR THE INVESTIGATION OF THE CHARACTERISTICS OF WATER 10

I. Physical Methods 10
1. Taking Water Samples 10
2. Measurement of Water Temperature 14
3. Measurement of the Intensity and Spectral Composition of Light 16
4. Measurement of the Rate of Flow of Running Water 18
5. Analysis of the Particle Size of the Substrate 19

II. Chemical Methods 20
1. Taking Water Samples 20
2. Determination of the Hydrogen Ion Concentration (pH Value) 21
3. Determination of Free Carbon Dioxide in Water (CO_2) 22
4. Determination of the Oxygen Content of Water 23
5. Determination of Ammonium in Water 27
6. Determination of Water-soluble Phosphate 29
7. Determination of the Alkalinity (SBV) and the Hardness due to Carbonate (Combined Carbon Dioxide) 30
8. Determination of the Hardness due to Calcium 30
9. Rapid Method for the Estimation of the Total Hardness of Water 31
10. Estimation of the Potassium Permanganate Consumption (Approximate Method) 31

CHAPTER 2. METHODS FOR THE INVESTIGATION OF THE OPEN WATER ZONE OF STANDING WATER (PELAGIAL) 33

A. Methods for the Investigation of Plankton 33

Zooplankton 34

I. Direct Observation and Recording of Zooplankton in the Water 34

II. Collection of Zooplankton from Water 37
1. Collection with Plankton Nets 37
2. Collection of Plankton Samples with the Aid of Water-bottles 46
3. Collection of Plankton Samples with Water pumps 47

III. Fixation and Preservation of Zooplankton 48

IV. Enumeration of the Zooplankton 48
1. Preparation of a Suitable Plankton Concentration and Taking Samples for Counting 49
2. Filling the Counting Chamber with Plankton 49
3. Counting the Plankton 51
4. Estimation of the Plankton Content of the Whole Sample 52

Phytoplankton 52

I. Qualitative Investigation of the (Net) Phytoplankton 55

II. Quantitative Investigation of the Phytoplankton 55
1. Fixation and Preservation 58
2. Counting the Phytoplankton 59
3. Other Methods of Evaluation 63

B. Investigation of the Living Community on the Surface Film (Neuston) 67

CHAPTER 3. INVESTIGATION OF THE BOTTOM ZONE OF STANDING WATER (LITTORAL AND PROFUNDAL) 69

I. Introduction 69

II. Investigation of the Eulittoral Zone 71
1. The Interstitial Ground Water of the Shore 71
2. Investigation of the Surf Zone on the Shore 72

III. Investigation of the True Littoral Zone (Infralittoral) 75
1. Reedswamp Zone (Upper Infralittoral) 75
2. Zone of Floating Leaves and *Potamogeton* Zone (Middle Infralittoral) 77
3. The Submerged Plants (Underwater Meadows) (Under Infralittoral) 80
4. Semi-experimental Methods for the Study of the Periphyton (Aufwuchs) 81

IV. Investigation of the Deep (Profundal) Zone 84
1. Bottom Grabs 85
2. Mud-borers, Core-samplers, and Mud-samplers 92
3. Dredges 93
4. Further Treatment of the Bottom Samples 95

V. The Quantitative Study of the Imagines Emerging in the Benthal Region 102

CHAPTER 4. METHODS FOR THE INVESTIGATION OF RUNNING WATER 104

I. Investigation of the Plant Population 106

II. Investigation of the Animal Population 109
1. The Population of Plant Mats 111
2. The Population of the Stones 111
3. The Population of the Finer Sediment 112
4. The Hyporheic Fauna 118

III. Investigation of Plankton in Running Water 121

IV. Determination of Organismal Drift 121

V. Collection of Insects from Running Water 122

VI. Semi-experimental Methods for the Determination of the Population of Running Water 123

VII. Further Treatment of the Samples 125

VIII. Presentation of the Results 126

CHAPTER 5. METHODS FOR THE BIOLOGICAL INVESTIGATION OF UNDERGROUND WATER 129

Investigation of Ground Water in Porous Rocks (Phreatic Fauna) 131

CHAPTER 6. METHODS FOR THE DETERMINATION OF PRODUCTION IN WATER 134

I. Determination of the Biomass 135
1. Chlorophyll Methods 136
2. The Dry Weight Determination Method 136
3. The Nitrogen Method 137

II. Determination of the Primary Production 137
1. The Oxygen Method (Gran Method) 137
2. The C^{14} Method of Steemann Nielsen (1952) 138

III. Some Remarks about the Determination of the Primary Consumption 145

IV. Population Dynamics of the Zooplankton 146

CHAPTER 7. METHODS FOR THE BIOLOGICAL ESTIMATION OF WATER QUALITY 147

I. Ecological Methods 147
1. Procedures which Operate According to Saprobic Systems 148
2. The "Species Deficit" of Kothé 156
3. Methods for Distinguishing Self-purification 157

II. Physiological Methods 157

APPENDIX I. Methods for the Fixation and Preservation of Groups of Freshwater Organisms 159
A. Animal Groups 159
B. Plant Groups 161

APPENDIX II. Some Remarks about Culture Methods 162
The Culture of Algae 164

APPENDIX III. Firms which Make Hydrobiological Apparatus 168

REFERENCES 169

INDEX 193

PREFACE

HYDROBIOLOGY is not a discipline on its own account, but is that part of biology which is concerned with the life of organisms in water. At the same time it is a special section of limnology, the science of inland waters. The objects studied by hydrobiology are animals, plants, and bacteria so far as they live in water. Hydrobiology therefore uses nearly all the methods generally used in biology. Special methods are those which make qualitative and quantitative collections of aquatic organisms possible. This book is concerned only with these specific hydrobiological methods. The methods of bacteriology have diverged in such a way from the divisions of hydrobiology—hydrobacteriology, hydrobotany, hydrozoology—that they require a separate description. The same is true of fishery biology, which has also developed an extensive methodology of its own. Finally, it is also to be noted that the methods of marine biology will not be considered in this volume. On the other hand, some of the methods of measurement and calculation used by production biology, which are today of the first importance in both hydrobiology and limnology, are included. Also considered are some simple physical and chemical methods which enable the hydrobiologist who does not contemplate a special analysis of the water to make a rapid and adequate estimate of it as an environment of organisms. This is important for hydrobiological work because its primary purpose is not the collection and determination of aquatic organisms but to gain an insight into the relationships between organisms and environment.

Efficient methods and clear questions are the foundations of scientific work. Experience shows that the ideas develop in so far as one "lives" with a problem. The methods are aids to the formulation of a question in such a manner that a clear answer can be expected. The methods must therefore be directed as accurately as possible to the special problem. Herein lies the difficulty of describing the methodological aids of a science, which is, as in the present instance, largely ecological and is, in its practical work, concerned with investigations in the field. All the methods cannot be described in detail. Many are not even mentioned. In order to give the reader further scope, numerous references to the literature are given in the text and also an extensive list of references. In this way it is also easy to find descriptions of special methods, and it seems to me that the addition of this list, selected for its practical value, serves as well as would the description of the methods themselves.

I am grateful to numerous professional colleagues for suggestions and materials for the book. I am especially grateful to my collaborators Frau Hildegard Backhaus and Fräulein Gisela C. Tillmanns for their careful preparation of the figures and photographs and also to my wife for her generous and continual help and collaboration.

Falkau im Schwarzwald JÜRGEN SCHWOERBEL

SHORT INTRODUCTION TO THE HYDROBIOLOGY OF FRESH WATER

I. TYPES OF WATER

Fresh water, in comparison with the gigantic marine waters, occurs as small and very small inland waters. Rivers and lakes are the most striking collections of water, and to these must be added underground water.

The lakes are of very different sizes and depths. The Feldsee in the Black Forest is only 380 m long and 32 m deep; the Bodensee is about 63 km long with a maximum depth of 252 m; Lake Tanganyika is 650 km long and 1450 m deep. Shallow lakes, the bottom of which may be everywhere colonized by plants, are called pools or—if they have an artificial outlet—ponds. Funnel-shaped cavities full of water, marshes, bogs on moors, pools in woods and meadows, and also drains (moats) are examples of small collections of water. Plant waters are also very interesting, being cavities in plants filled with water as, for example, the cavities in beech-tree trunks. Collections of water which persist for only a short time, but recur more or less regularly, are periodic or temporary waters.

In contrast to a lake a river is not delineated from the land on all sides but is open "above and below". The regular current determines the removal of all substances dissolved or not dissolved in the water. Rivers have a dimension more extensive than lakes, the longest river, the Missouri–Mississippi, being 6790 km long. The rapidly flowing brooks of the mountains are distinguished from slowly running rivers. Every river, every brook, arises from a spring, ground water escaping from the earth with characteristic living conditions and organisms.

Underground water appears, apart from springs, under the surface of the earth in cavities as cavernous lakes and rivers and as underground water in the fissures in rocks and in the porous spaces of the sediments.

II. LEVELS OF CORRELATION

1. Water as an Environment for Organisms

Every organism is affected by factors in its environment. The temperature determines, within the physiological range, the speed and intensity of biochemical reactions. In animals increase of the intensity of respiration is especially important. Thus in the flatworms *Crenobia alpina* and *Dugesia gonocephala* the oxygen consumption first increases with increasing temperature and then falls again. The maximum oxygen consumption of *Crenobia* occurs at 15°C and that of *Dugesia* at 20°C (Schlieper). Important conclusions can be drawn from this fact about the occurrence and distribution of these two species, and about the competition between them. At higher temperatures most aquatic animals also move more quickly. The curves of the speed with which the named flatworms crawl coincide with those of their oxygen consumption, so that we can assume a direct relation between the two curves. The occurrence of the algae and of the higher aquatic plants is often related to a definite temperature range. In general there is, with increasing temperature, an increase of the production of material and of growth and turn-over of nutrients. In all the examples mentioned, to which others could be added at will, there always appears a maximal effect which declines again with a further increase of temperature. Another important factor is the oxygen content of the water. Decrease of this reduces the oxygen consumption of animals either proportionally or after a long compensation period. The lethal oxygen tension for animals in running water is as high as it is for those in still water (Ambühl, 1959). Similarly, the algae and mosses in running water seem to be dependent on a high oxygen partial pressure. The hydrogen ion concentration (pH value) influences the metabolism in both animals and plants. In detail it is nevertheless difficult to determine the specific effects of this factor. It is self-evident also that mineral nutrients, such as phosphate and nitrate, as well as iron and other elements, are of direct significance. Algae, for example, can take up in a very short time 30 times their phosphate requirement from the water and store it up (Grim).

The ionic composition of the water plays a decisive part in the presence or absence of plants and animals in water. Freshwater animals live in a hypotonic solution, i.e. the water is always poorer in ions than their body fluids are, and the problem for the animal is the constant maintenance of the concentrations of certain ions in its body. Further, light is also among the factors that operate directly on organisms, and its importance for assimilation by algae and also for the distribution of zooplankton will be discussed further below. For organisms in running water the water currents are very important (see p. 104) and, finally, the structure of the bottom is important.

2. Community Life in Water

The organisms are, however, not only exposed to the factors in the environment; they also mutually influence one another and form communities (biocoenoses). Thus an animal biocoenosis develops in standing littoral water which is temporarily parallel to the higher vegetation. The aquatic plants are food (e.g. for leaf miners), and form substrates on which eggs can be laid (by dragon flies, aquatic mites, caddis flies, etc.) and materials for the construction of cases (by the larvae of caddis flies, and aquatic butterflies, Fig. 1) and a supporting structure for other communities (*Aufwuchs*, periphyton).

The course of the life of an animal is always bound up with other animals in such a way that their relationships have the character of a mutual competition. The following diagram after Illies shows the conditions to which a species is subjected within the biocoenosis:

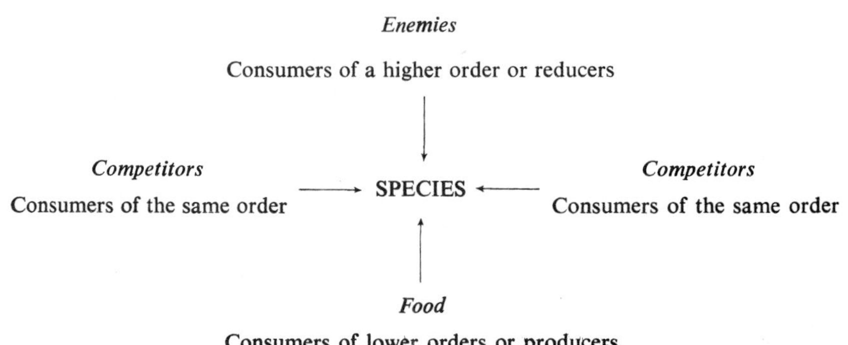

FIG. 1. Floating leaves of *Potamogeton natans*, out of which the caterpillar of the aquatic butterfly *Hydrocampa nymphaeoides* has cut pieces of leaf to make its case. The feeding tracks of leaf miners (chironomids) can also be seen.

3. Total Life in Fresh Water

The lakes also are not permanently static water; they undergo an annual rhythm of stagnation and circulation. In the summer a warm upper layer (epilimnion) lies over the cold deep layer (hypolimnion), with the result that the temperature in the transition zone (metalimnion) suddenly falls (thermocline). In the autumn the surface layer cools off. Storms lead to a mixing that penetrates even deeper until the equality of temperature in the lake causes a circulation of the whole watermass (autumnal circulation). In the winter the epilimnion then cools to the point of ice formation while the temperature in the deep zone remains constant at about 4°C (winter stagnation). In the spring, after the snow-melt, the warming of the upper layer begins until the temperature reaches 4°C in the whole lake, and this leads, with the co-operation of the wind, to the springtime full circulation. Further warming restores the typical three-layer system in the lake (summer stagnation).

This type holds for many mid-European lakes. The epilimnion is the most heavily populated layer; the trophogenic zone is predominantly the zone of primary production. Oxygen is always abundantly present in it because of exchange with the atmosphere and also the assimilation of the green plants. The hypolimnion, on the other hand, is shut off from the atmosphere, and during the stagnation period consumes the oxygen reserves brought to it by the full circulation. The consumption is twofold:

(1) By the putrefaction of dead animals and plants, which sink down in a constant rain

from the epilimnion: tropholytic zone. Putrefactive processes are oxidations which use up oxygen.

(2) By the respiration of the organisms living on the lake bottom.

The extent of the oxygen consumption at the bottom depends in the first place on the shape of the lake basin. This is easy to understand if one considers that the deeper a lake is the longer a dead daphnia, for example, lying in the epilimnion, takes to sink to the bottom. If putrefaction of the animal is limited to this sinking time, i.e. if the organic substance is completely oxidized to mineral constituents, only the empty shell of the daphnia reaches the lake bottom and this consumes practically no more oxygen. But for this reason the nutritional supply of the organisms living at the bottom of the lake is always meagre and the population density is correspondingly moderate. In shallow lakes the situation is quite different. The putrefaction of dead organisms proceeds intensively at the bottom of these lakes, and the bottom zone is, because of its high nutritional supply, very densely populated. This leads in the course of the stagnation period to a marked decrease of oxygen, even to a complete absence of oxygen from the depths. These conditions naturally influence directly the organisms living at the bottom, as Thienemann especially has shown.

In standing water a characteristic form of life has developed—that of the plankton organisms floating or drifting in water. Floating is a constant compensation for the rate of sinking. According to Wesenberg-Lund and W. Ostwald

$$\text{the rate of sinking} = \frac{\text{excess of weight}}{\text{form resistance} \times \text{the internal friction}}.$$

The internal friction of the water is a physical value which cannot be influenced by the animal. It depends on the viscosity of the water which diminishes with increasing temperature. At 0°C it is twice as much as at 25°C, so that a plankton organism would, when other conditions remain the same, at 25°C sink twice as quickly as at 0°C.

In order to keep the rate of sinking as low as possible, the excess of weight must be reduced and the form resistance increased. Study of the animal, and especially of the plant plankton, shows strikingly how this is achieved. The formation of gas and fat vacuoles and increase of the volume by means of gelatinous envelopes reduce the excess weight, and the form resistance is increased by extending the surface by means of spikes, spines, and other outgrowths (Figs. 2 and 3). In some zooplankters the body form undergoes a cyclical change. For example daphnias reproduce for many generations during the summer period parthenogenetically, i.e. without males. In some species and races the form changes, the head becoming longer and forming a "helmet" (Fig. 4). The tail spine may also be lengthened. Probably this form change has nothing to do with flotation, as was formerly believed. It can be produced experimentally by increasing the temperature, light, and turbulence and in rotifers also by special feeding; and it occurs only in plankton organisms which have considerable power of independent movement and not in those which actually only float. The possible advantage of better form resistance is far exceeded in these animals by their increased mobile activity in the summer.

One might suppose that plankton organisms floating in water, or not moving much in it, must be homogeneously distributed in the lake or could accumulate anywhere only incidentally. This is not in any way true. The chlorophyll-bearing algae, which depend on light energy, only live in the upper part of the lake in the trophogenic zone and extend down to variable depths according to the optical properties of the water. Plankton animals undertake many regular "vertical migrations" in that at the beginning of dusk in the evening

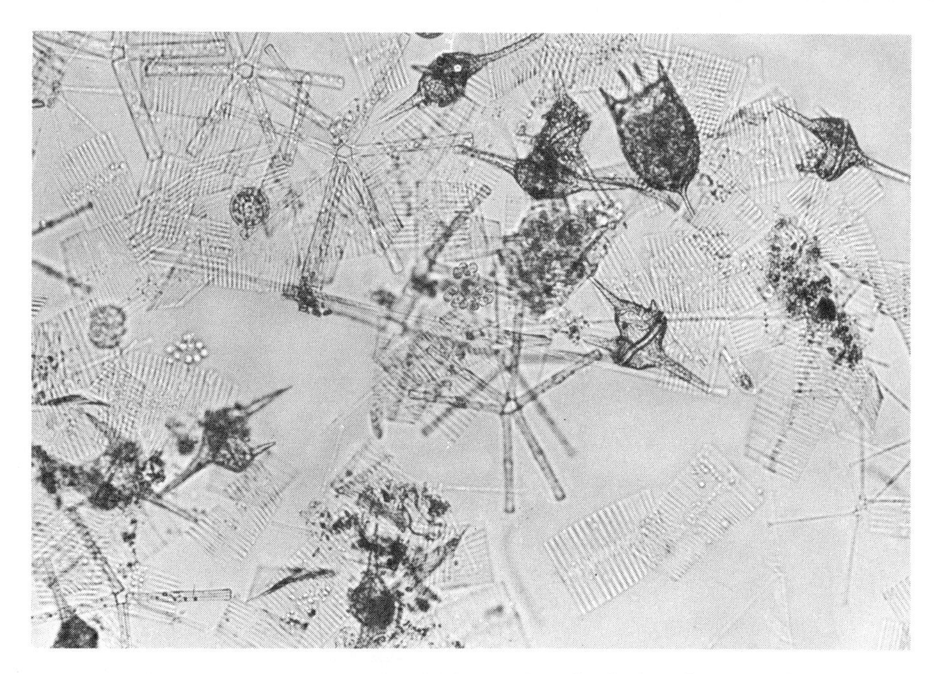

FIG. 2. Net plankton with (below on the right) star-shaped colonies of *Asterionella* and (in the middle of the figure) *Tabellaria*. On the right above are bands of *Fragilaria crotonensis*, *Ceratium*, and *Keratella*.

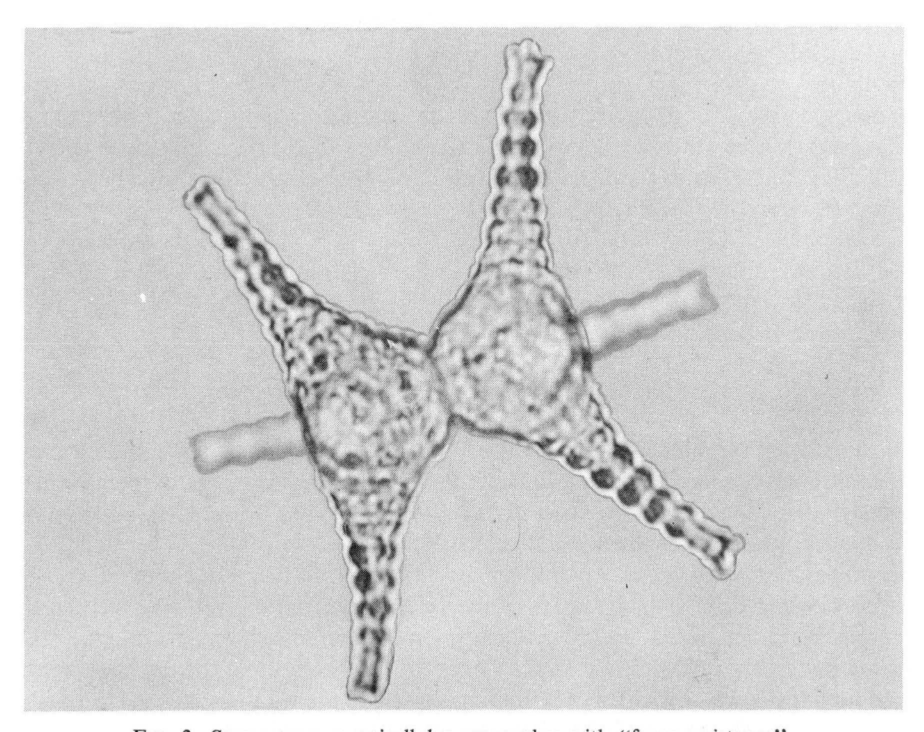

FIG. 3. *Staurastrum*, a unicellular green alga with "form resistance".

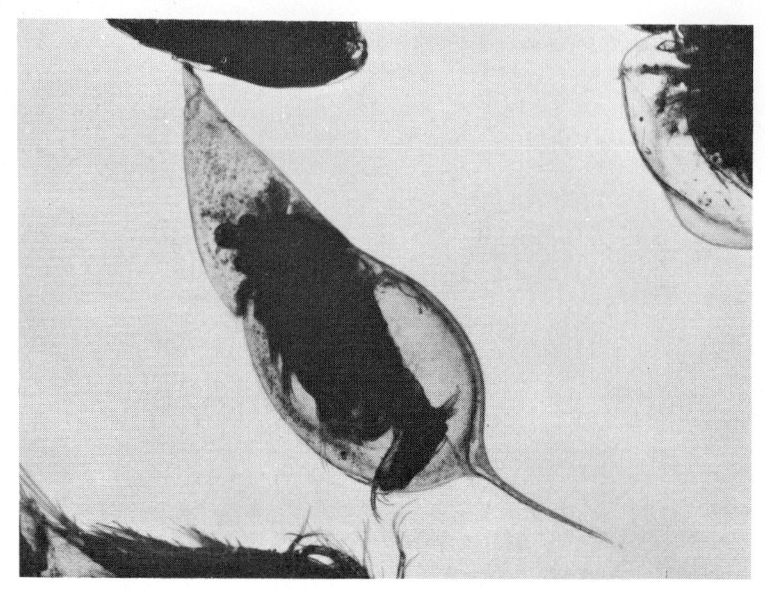

FIG. 4. Summer daphnias with a tall helmet, from the plankton of Lake Balaton.

they move upwards from the deeper layers to the surface and in the morning return to a certain depth in which they remain for the rest of the day. As recent investigations have shown, this behaviour is in the first place controlled by the change in the light intensity in the morning and evening. In stagnant water there are, naturally, not only plankton organisms. Fish, for example, swim about freely in it; they are called nekton in contrast to the plankton.

The living community of the bottom zone, the benthos, is composed of the plants and animals of the shore (littoral) zone and the animals which live in the deep (profundal) zone, of the lake. Very interesting is the neuston, organisms which are attached to the surface film or live on the surface of the water, e.g. water-skaters.

Once more let us return to the algae. These are the organisms which synthesize the first formed organic substance from light energy and the nutrient salts dissolved in the water. They are, as it were, the original or primary producers in the water. They require for this photosynthesis, according to the equation

$$6\,CO_2 + 6\,H_2O \xrightarrow{h\nu} C_6H_{12}O_6 + 6\,O_2,$$

carbon dioxide as supplier of the carbon which is the most important building block for organic syntheses in the plant. The photosynthetic production of algae in the lake has been investigated with the aid of radioactive carbon (for the method, see p. 138), and values have been obtained which are probably characteristic of the various lakes.

The whole animal world in the water lives on this primary production. First come the filter-feeders which directly filter out the unicellular algae from the water and especially the very minute nannoplankton. Daphnias have been cultivated together with algae "tagged" by the intake of radioactive carbon. After a time the radioactivity of the daphnias was measured and from this it was calculated how many algae the animals had eaten during this time and how much water had thus been filtered. The results surprisingly showed that in spite of intensive filtration activity only 5% of the supply of algae were used up. This result relates to a certain lake in Sweden, but the situation is probably similar in other lakes about which there are so far no thorough investigations. A quite important part of the food of the filter-feeders is bacteria as well as algae, and also inanimate organic particles. The filter-feeders are in any event the first (primary consumers) to pass on the primary production to the subsequent producers and on these live the predators and ultimately the fish (Fig. 5).

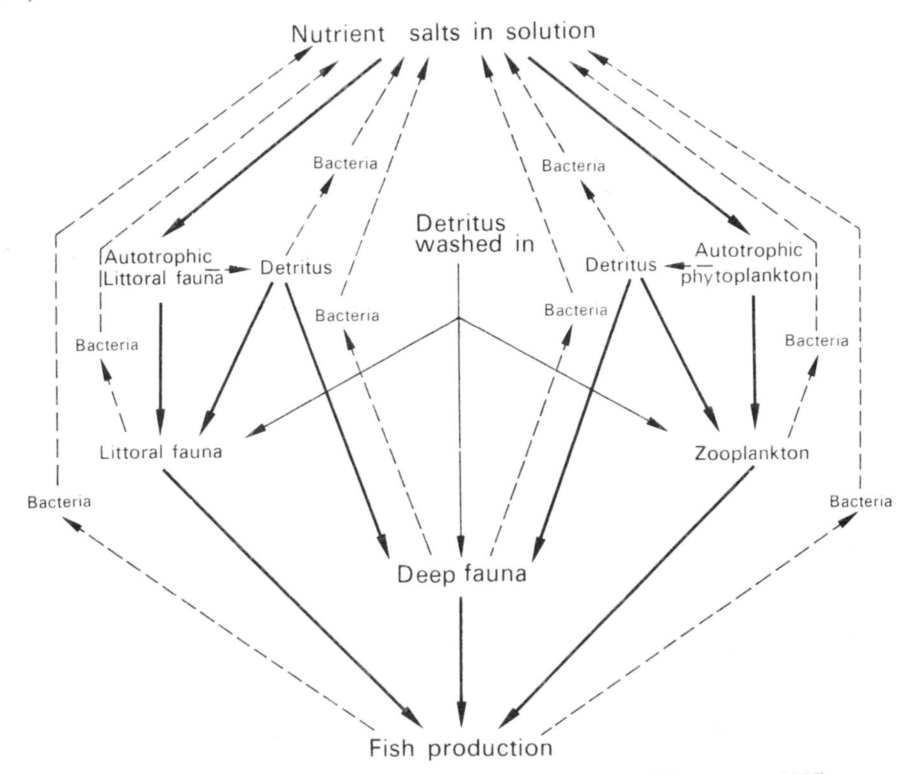

FIG. 5. Diagram of the metabolic cycle in a lake (from Thienemann, 1955).

The substances return to the water by bacterial breakdown of the dead organisms or else the bottom animals live on those of them which reach the bottom of the lake in the form of organic substances. The variations in the various lake types which result are mentioned on pp. 84 ff.

A river is, in contrast to a lake, an "open system"; there is no circulation of water in it, but a directed transport. Further, the river is not organized vertically, but horizontally. In it all the factors in the environment change from the source downstream. The currents and gradients decrease gradually, the sediment in the bottom is correspondingly always more finely granular, and the summer temperatures increase downstream. The intake of oxygen is, when there is more marked turbulence, so great that there is a considerable oxygen deficit only in sluggish streams and only in zones directly affected by decomposition. This occurs where there are large discharges of organic sewage into the river. In the region directly affected by discharge there occur only decomposition processes that use up oxygen (the *poly-saprobic zone*). However, when there is a good oxygen intake, the mineralization of the organic substances soon sets in *(meso-saprobic zone)*. This "self-cleansing" of the running water is a process that is in the highest degree characteristic and is also of practical importance. It leads from the poly-saprobic zone through the meso-saprobic zone to the oligo-saprobic zone with only small residues of the pollution. In each of these zones there is a characteristic living community.

Every organism in a river lives where it finds especially good living conditions. But the mechanical pull of the water current creates a constant risk that it may be drawn into other environmental conditions. It is necessary to guard against this.

Let us consider the mountain stream. Plankton organisms are completely absent from this. The nekton is represented only by some species of fish that have a characteristic form and mode of life. The neuston is also, for obvious reasons, absent. The most important difference from the lake is the fact that almost the whole of the animal and plant population of this environment belongs to the benthos. As a protection against being torn away the plants root themselves on the bottom or attach themselves to rocks or other substrates, as the algae do. Only a few of the animals live in the strongest currents. They hold on to rocks by means of suckers (the larvae of the midge, *Liponeura*) or they press their flattened bodies close to the substrate (e.g. the larvae of some may-flies and stone-flies); some cling to the substrate (e.g. the larvae of some black-flies and the pupae of *Liponeura*), but most of the animals seek to escape from the immediate current by living between the rocks where the current is much weaker. For this reason the cushions of moss are often densely populated: species which live here hold on to the branches of the plant with especially strong claws and spines. A great many animals live in the river sediment itself; they either dig out their own tunnels (e.g. the larvae of some may-flies) or make use of pore spaces arising from the open arrangement of the irregularly formed granules (the hyporheic sphere of life or hyporheal).

The further we go downstream, the slower the current becomes and the living conditions so much the more resemble those in the lake. Plant and animal plankton appear, the number of species of fish increases, and other organisms found in stagnant water populate the river. This is especially true of the artificial and natural reservoirs on the stream.

The supply of energy and food is, compared with the volume of water, greater in the river than in the lake. The shore and bottom plants are of much greater importance as primary producers than they are in stagnant waters (excepting ponds, pools, and small collections of water). Here are found the *Aufwuchs* algae as well as, in the larger rivers, the plankton algae. Consumers are the filter-feeders which assemble especially under the zones of still water, and also the detritus feeders are predators.

When we mention the hyporheic environment we have already taken a step to the underground water. Subterranean environments are shut off from the surface of the earth and are characterized by the absence of light. In addition the temperature of the water is more or less constantly low in so far as we know it from the springs. The hyporheic environment constitutes a transition region with conditions of life similar to those of the river, but with complete darkness, little movement of water, and temperatures the daily and annual variations of which are much more equalized.

Because of the absence of light, underground waters are populated almost entirely by animals which, in comparison with species living above ground, show a number of characteristic features. Many are blind or have reduced eyes, and avoid light: in addition the body pigment has to a large extent degenerated. On the other hand, many of them, in so far as they live exclusively in cave waters, have very long feelers and legs with which they make contact with their environment.

The true underground water organisms, which live in the underground stream of the river shingle and remain in the interstices of this, are mostly markedly elongated and thin. The legs and the other appendages of the body are shortened and to some extent reduced, and everything is organized towards the body shape of a worm. These animals move in a snake-like manner through the narrow spaces. This type is already found in the hyporheic sphere of life in which its development from the organisms living above ground in streams can be followed.

CHAPTER 1

SOME PHYSICAL AND CHEMICAL METHODS FOR THE INVESTIGATION OF THE CHARACTERISTICS OF WATER

EVERY kind of hydrobiological work involves at the same time an investigation of at least the most important properties of water. Otherwise a biological comparison between different regions of an expanse of water, or between waters of different types, is not possible without an *a priori* abandonment of the causal analysis of the biological differences. Hydrobiology can, above all, forego investigations which were important for the assessment of special limnological problems, e.g. the measurement of currents in stagnant waters (not in running water), and also perhaps the determination of the phosphorus components, the nitrites, silicates, etc., in water. For certain hydrobiological problems the accurate knowledge of these conditions is of fundamental importance; the currents in a lake, for example, for the distribution of plankton, the silicate content of the water for the development of diatoms.

Accurate knowledge of all the chemical components and of the turn-over of these in the water and in the organisms is the basis of the study of the metabolism of the water. Considerable use of apparatus is nowadays made to carry out these investigations.

Therefore this book will describe only some of the measurements and chemical methods constantly in use which can be carried out by non-professional hydrobiologists, by students, or by the study groups of higher schools which are interested in hydrobiology. The numerous references to the literature show those who are interested where they can find special methods. For the methods of water analysis there is a series of good descriptions to which reference can be made when necessary.

I. PHYSICAL METHODS

1. TAKING WATER SAMPLES

There is no difficulty in taking samples of water from small or shallow waters, easily approachable from the shore, such as shallow pools and ditches, or the marginal areas of larger pieces of water and the surfaces of these (by boat) or even from small brooks and springs. The water sample is taken directly from these, and for this purpose polyethylene bottles serve especially well; they are lighter than glass bottles and are unbreakable. So far as it is known they do not give off substances into the water but, nevertheless, absorb substances from it. This is prevented if the polyethylene bottles are iodized for a week. To do this they are filled up to the brim with a 5% solution of iodine in an 8% solution of potassium iodide. During this period the free nitrogen valencies of the polyethylene are saturated with iodine. The bottles are finally washed out thoroughly with distilled water. The iodine solution can be used again (Heron, 1962). Polyethylene bottles are also cheaper than suitable glass bottles of the same size. If, however, glass bottles are used, they should be used exclusively for this purpose and remain longer in use. Naturally each of them will be carefully washed out before the sample of water is taken. Glass bottles do not give off any substances into the water if their internal surfaces have been roughened with hydrofluoric acid. To do this treat them with 0·5% hydrofluoric acid in distilled water for 8 days (report by Hassenteufel *et al.* (1963) for Pyrex glass vessels, which compare well with Jena apparatus Glass G20).

At greater depths, water samples must be taken with the aid of sampler bottles. The simplest and most easily produced of these is Meyer's sampler bottle (Fig. 6). It is best to use

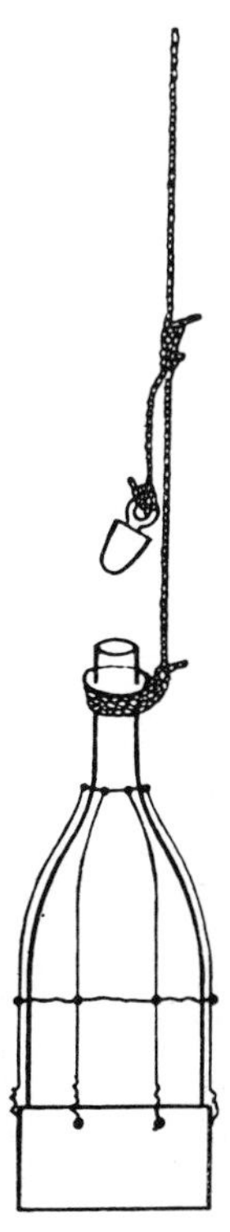

Fig. 6. Meyer's sampler bottle. (After Ruttner.)

an empty wine bottle enclosed in a net of string or cord and weighted below with a stone or lead weight. A strong cord is attached to the neck of the bottle to which the cork is so attached that it does not reach the neck of the flask when the cord is tightly stretched. The corked-up bottle is let down to the desired depth and is then opened by a strong pull on the cord. When the bottle is full it naturally cannot be corked up again and must be pulled up open, so that admixtures with other layers of the water cannot be avoided. One must therefore understand clearly that this method is not wholly free from error. Nevertheless, it is still a simple aid of the greatest value. The bottle can only be used to a depth of 50 m because at greater depths the hydrostatic pressure is too great.

The special, complicated (and more expensive) water-bottles follow a principle the reverse

(a)

(b)

FIG. 7. Ruttner's water-bottle: (a) open, (b) closed.

of that of the bottle just described. They are open above and below and are let down open to the desired depth and are closed by a drop-weight and a spring mechanism. In this way a sample of water is isolated which is not affected when the bottle is pulled up through other layers of water. The water sampler of Ruttner (Figs. 7a, b and 8) is especially good. It consists of a plexiglass cylinder which can be closed above and below by horizontal ground-in covers. When the cylinder is let down the covers are open so that the water can stream through the cylinder (Fig. 7a). This causes eddies to develop on the horizontal covers which cause disturbances if narrow, limited, vertical layers in the water are to be investigated. For plankton studies, therefore, one uses other models, described on pp. 46 ff. If the bottle is to be closed, a drop-weight weighing 0·5–1 kg is allowed to fall down

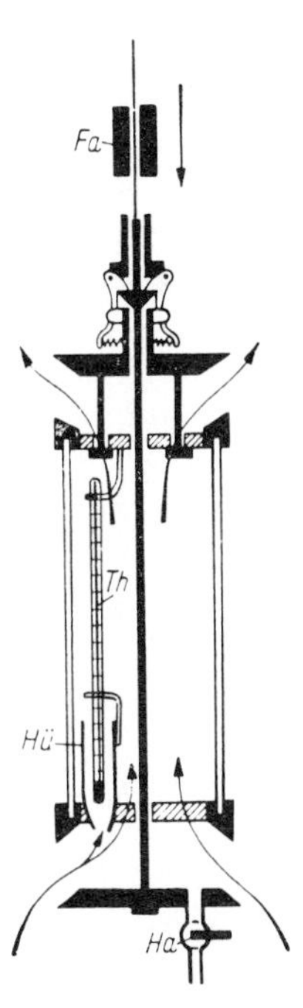

FIG. 8. Longitudinal section of Ruttner's water-bottle. *Fa*, the drop-weight (messenger); *Hü*, the protective cover for the thermometer, *Th*; *Ha*, the drainage cock (see text). (From Ruttner, *Fundamentals of Limnology*.)

FIG. 9. Drop-weight (messenger) for closing limnological apparatus under water. (Photo by Hydrobios.)

the line on which the cylinder hangs (Fig. 8), and this strikes a spring mechanism which releases the device which closes the bottle. The glass cylinder falls on to the lower cover and the upper cover almost simultaneously falls over the upper opening. In this way the bottle is closed and can be drawn up. Inside the glass cylinder there is, if it is appropriate, a small thermometer which indicates the temperature at particular depths. In the lower cover a drain cock is mounted with a tubular connection, so that the cylinder can take as many samples as are desired.

Perforated lead, iron, or brass cylinders serve as drop-weights (Fig. 9), and these are commercially available. To suspend the apparatus, perlon or nylon cords are best. For heavy apparatus, also used for greater depths (50 m and deeper), one cannot manage without a wire cable and a cable-winch. It is a great advantage to prepare the perlon or nylon cord before use to show intervals of 1, 5, and 10 m. These can be marked with thread of different colours twisted in firmly several times at the measured points (see Fig. 55, p. 86). Different colours are chosen for 1, 5, and 10 m. Care must be taken that these markers do not slip and that the drop-weight is readily movable on the line. The markers should be checked from time to time.

For taking water samples from running water, Jaag, Ambühl, and Zimmermann (1956) developed a horizontal water-bottle. The cylinder is not let down vertically into the water, as the Ruttner cylinder is, but horizontally. The covers on both sides of the cylinder are kept open by springs and can be closed by traction on two cables. Wohlstedt and Schweder had earlier developed horizontal water-bottles. Joris (1964) briefly described a horizontal water-bottle for taking water samples at the bottom (see also pp. 92 ff).

2. Measurement of Water Temperature

An accurate knowledge of the temperature conditions in a water-body is always important in hydrobiological investigations. Basically three procedures are available for this. Measurement of the temperature with mercury and fluid thermometers, electrical measurement of the temperature, and chemical methods of measuring it.

The simplest methods are measurement with mercury and fluid thermometers. These can be put into the water directly from the shore. The maximum–minimum thermometer is very useful; it records the extreme values at any desired period of time. Measurements of this kind are especially interesting in springs and streams and in general in all small collections of water which show marked daily temperature variations, and also in the hyporheic ground water. For taking the temperature at different depths in lakes and the larger rivers one can also use the water samples taken in water-bottles, these being, because of the high heat-capacity of water, sufficiently accurate. The thermometer built into the water-bottle acts in the same way.

Formerly, the tilting, reversing thermometer (Fig. 10) was used exclusively for accurate determinations of the vertical temperature stratification in lakes. This thermometer swings inside a metal frame. It is let down to the desired depth on a marked cord and exposed there for about 5 min. The apparatus then shows accurately the temperature at this place in this layer of water. In order to fix this temperature on the spot a drop-weight is let down and by means of a spring mechanism the thermometer is released from its frame—it tilts through 180 deg and this pulls the thread of mercury down to a narrowing of the capillary tube and guides it to the other—now the lower—end of the thermometer which is graduated in 0·1 deg. The temperature in the upper part of the instrument can no longer affect the length of the mercury thread, so that the temperature of the depth investigated can be read off at the surface with great accuracy. Reversing thermometers are still on the market

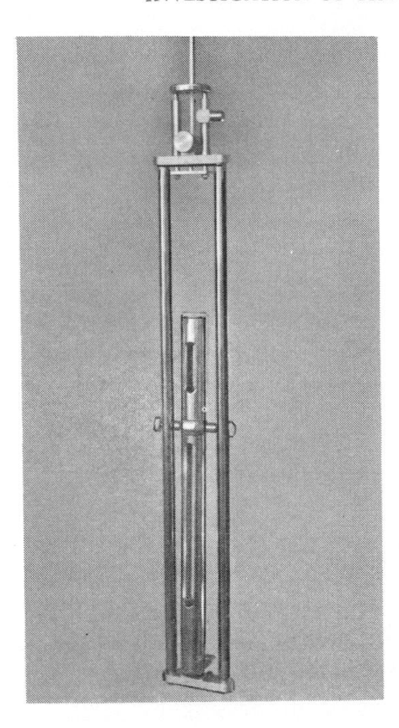

FIG. 10. Reversing thermometer. (Photo by Hydrobios.)

but, because of the great precision and requisite calibration of the thermometer, they are rather costly and therefore no longer competitive in comparison with devices for the electric measurement of temperature.

Nowadays the reversing thermometer has been superseded by other methods not only in order to economize time but also for technical reasons. Every measurement made with the reversing thermometer takes 6–20 min according to the depth. To make a vertical temperature profile in deep lakes one requires several hours. But accurate investigations have shown that the vertical temperature relationships in lakes may be altered considerably in a very short time a by shifting of the isotherms. The reversing thermometer is too cumbersome to measure these changes, and a change to electrical methods of measuring temperature has been essential to methodological development. These methods do not give more accurate results, but they record them without delay and that is the important thing.

Now in use are thermocouple elements or an electrical resistance thermometer, into the construction of which it is not necessary to enter here. The recording apparatus remains in the boat while the temperature probe is sent down to the depths. Because the instruments are, during their operation, almost free from inertia, a temperature profile can be taken with these in only a few minutes. It is also possible to record the smallest variations, especially near the surface and within the thermocline which formerly could not be done with the necessary accuracy because of the size of the reversing thermometer. Of course, the electrical temperature can be connected to an ink-recorder so that a continuous record of the temperature can be taken. Mortimer (1953), Mortimer and Moore (1953), and Schmitz (1954) have summarized the methods used in hydrobiology.

Finally, reference may be made to a method of Pallmann, Eichelberger, and Hasler (1940) by which the temperature is measured for a longer period of time indirectly by its effect on the speed of chemical reactions. The reaction used for this is the inversion of sucrose to invert sugar (glucose and fructose). The inversion of sucrose depends on time and temperature. The different abilities of the kinds of sugar to rotate light are easily determined by means of a polarimeter (see Schmitz (1954) and Schmitz and Volkert (1959)). The test solutions are fused into glass ampoules and can be exposed for weeks or months.

Conversions for degrees Fahrenheit, Reamur, and Centigrade:

$$t°\text{C} = \tfrac{4}{5}\times t°\text{R} = (\tfrac{9}{5}\times t+32)°\text{F},$$

$$t°\text{F} = \tfrac{4}{9}\times(t-32)°\text{R} = \tfrac{5}{9}\times(t-32)°\text{C},$$

or 5 degrees Centigrade = 9 degrees Fahrenheit = 4 degrees Reamur.

3. Measurement of the Intensity and Spectral Composition of Light

During recent years the methodology of measuring radiation under water has been very much improved. In this it is a question of measuring the intensities and qualities of light and also the turbidity.

The light climate in water is of especially great importance for the photosynthesis of plants and thus for primary production (cf. pp. 78 and 134. ff.) as well as for the distribution of plankton in the water. Measurements of radiation are therefore also important for hydrobiological problems, but they cannot be carried out with simple aids.

The apparatus used operates like the illumination meter of a photographic apparatus. Barrier-layer photocells generate, when they are illuminated an electric current proportion-

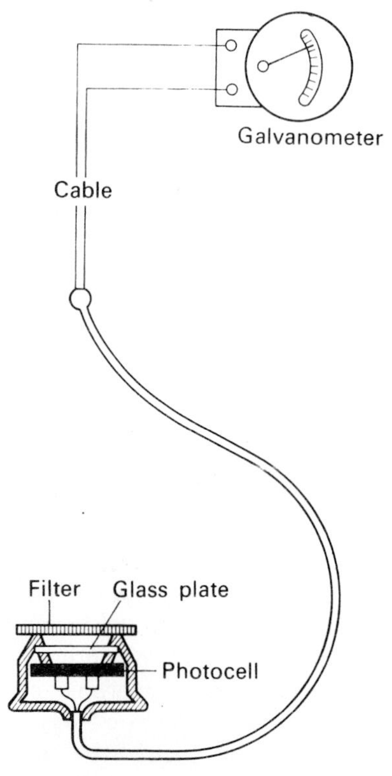

FIG. 11. Simple apparatus (barrier-layer photoelectric cell) for measuring light under water. (After Eckel and Ruttner, from Ruttner, *Fundamentals of Limnology*.)

al to the strength of the illumination and these currents can be measured by a galvanometer (Fig. 11). Naturally the photocell must be housed in a waterproof container. It is exposed in the water on a cable, the measuring instrument being kept in the boat or on the shore. If one wants to measure the spectral composition of the light under water as well as its intensity, a colour filter of known transmission and colour quality (wavelength) is put in front of the photocell, and one determines the intensity of the individual parts of the spectrum. The scale of the galvanometer can be directly calibrated in lux.†

As a method for estimating the decrease of light intensity in the water, the depth of visibility is in general determined by the Secchi disc (Fig. 12). This is a white disc, circular or quadrangular, with a diameter of 20–25 cm, which is allowed to sink down slowly on a marked line until its outlines just disappear. The depth at which this happens is the depth of visibility and this is a measure of the transparency of the water. It is, for example, very much affected by the plankton content of the water.

This simple, easily made apparatus serves well on excursions for making preliminary guiding measurements.

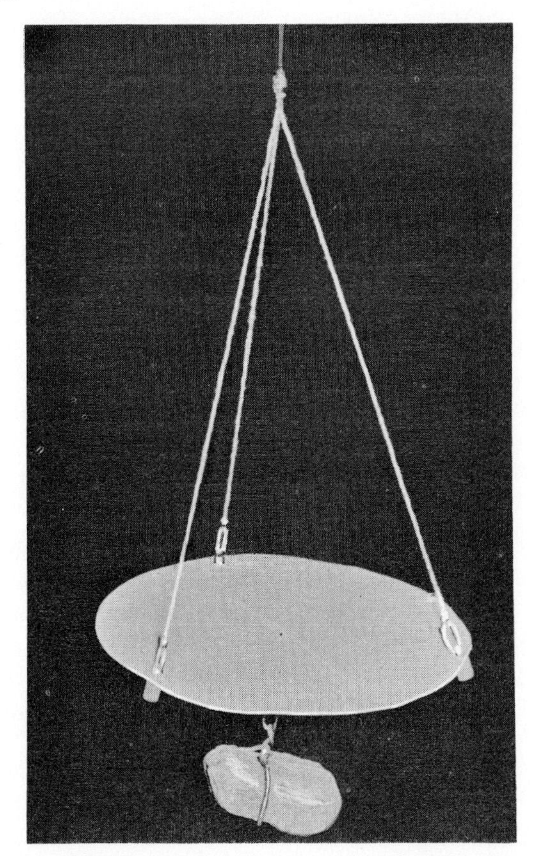

FIG. 12. The Secchi disc for measuring the depth of visibility.

† The reading of American works is more difficult because in them the light intensity is stated in foot-candles: 1 foot-candle = 10·83 lux. Because a detailed description of other methods would take up a great deal of space, reference will be made here to the following readily available literature: Görlich (1951), Sauberer (1962), Sauberer and Eckel (1938), Sauberer and Ruttner (1941), and Schmitz (1953 and 1960).

4. Measurement of the Rate of Flow of Running Water

Even in an ideal artificial channel the speed of the flow of water varies at any point on the transverse section of the channel and is least in the immediate neighbourhood of the solid bottom. The layer of water near the bottom is, in natural running water, enlarged to an extraordinary degree by rocks and other irregularities in the substrate and its structure is complex. Because this layer of water is biologically the most important, it is especially regrettable that there is so far no method of accurately measuring the water movement here.

Colour method

There is, on the other hand, a simple subjective method, which at least makes the movement of water on the bottom visible. The water is carefully coloured. Crystals of potassium permanganate or other substances that are strongly coloured and are readily soluble in water are placed on the bottom in glass tubes open at both ends or with both ends drawn out to a point, and the direction of the coloured stream of water that emerges is directly followed. Its diffusion and the movement of the water can be measured with a stop-watch with some degree of accuracy. This method gives important indications in the study of the distribution of organisms at the bottom of small streams.

Drift methods

An object floating on the surface of a stream is carried along at a speed corresponding to that of the water current. If one records the time required for its transport between two fixed points the speed of the current can be calculated. According to Uhle (1925) the average speed of the water amounts to about 85% of this surface value. Naturally this speed does not correspond to the actual speed of the water at the places in which the organisms live, but the average speed of flow works out according to the configuration of the bed of the stream and to the composition of this bed and the species and number of the higher plants growing on it and also to the biological conditions; it should therefore always be ascertained.

Other methods

In addition hydrometric propeller-vane meters are used, the number of revolutions of which per unit of time depends on the speed of the current. The smaller the vane is, the more accurately can one sense and investigate the cross-sections of the water, but still one does not in this manner make contact with the bottom layer of water. In slowly flowing streams and in very turbulent water the vanes give values that are too low (Engelhardt, 1951; Zahner, 1959). Richardson (1929) was able to reach with the hot-wire instrument the biologically interesting layer, but these instruments are too sensitive for use in the open. Two other instruments, the Petit tube and the Gessner funnel, utilize the hydrostatic pressure of the flowing waves. With the Petit tube the impact pressure is directly measured manometrically, but the marked variations give results that are too inaccurate (Stuart, 1953, cited by Zahner). Gessner (1950) measure the amount of water collected during a certain time by a small bag provided with an inverted funnel-shaped opening which is held against the stream. The instrument must naturally be calibrated before use at known speeds of flow.

5. Analysis of the Particle Size of the Substrate

The composition of the substratum is important both for the population of the more stagnant and the running water of the bottom zone and of the hyporheic as well as the interstitial ground water. Analysis of the sediment according to particle size is therefore necessary for biological problems.

Technique

The sample of the substrate is taken from the water-body (for the technique of this, see the special sections) and it is then dried. The particles of different sizes are then separated from one another by means of sets of standardized sieves arranged above one another according to the sizes of their meshes (Fig. 13). The sample to be sieved is poured on to the uppermost, coarsest sieve, and the whole set of sieves in enclosed in a cover and put on to a shaking machine.

After 15 min the individual fractions are taken out and weighed.

Fig. 13. Set of sieves for the analysis of the size of particles in samples of sediment.

Interpretation of the results

The percentage weights are assembled in a table and are later expressed in a semi-logarithmic cumulative curve (Fig. 14). The curve follows a course which corresponds to the quotas of the individual granules. When the curve rises steeply the quota of the corresponding particle sizes is high, and when the curve flattens, it is low. The particle sizes at 10, 50, and 60% of the whole sample read off from the curve are especially important. Fifty per cent corresponds to the average particle size, the value 60/10 indicates the heterogeneity of the sample of sediment. Figure 14 shows how different this value can turn out to be in different samples of sediment. It stands to reason that this state of affairs operates directly

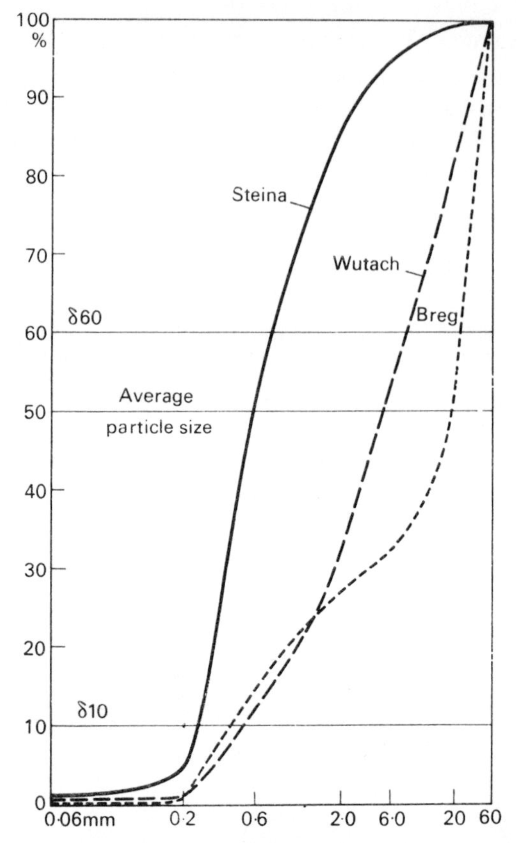

FIG. 14. Summation curve of the fractions of particle sizes.

on the bottom fauna. The following table shows the customary designations of the classes of particle size:

Stones	over 63·0 mm	Coarse sand	2·0–0·63 mm
Coarse gravel	63·0–20·0 mm	Medium sand	0·6–0·2 mm
Medium gravel	20·0– 6·3 mm	Fine sand	0·2–0·063 mm
Fine gravel	6·3– 2·0 mm	Silt	below 0·063 mm

II. CHEMICAL METHODS

The simple techniques here given are not intended for a thorough chemical analysis of the water, but are field methods for the on-the-spot examination of water and laboratory methods which can be carried out with simple equipment and nevertheless give good results. Because there are a number of readily accessible descriptions of the chemical analysis of water, only the most important methods will be described here.

1. TAKING WATER SAMPLES

The most important of the methods of doing this have already been described (see p. 10). In general 1 l. of water is enough for chemical investigation. From springs, streamlets, and ground water (see p. 131) one collects the water first of all in small flasks and decants it into a 1-litre flask. Each flask is given a number which is entered in the notebook which records the kind of water, the place where the sample was taken (on the shore, between plants, in the middle of the lake, etc.), and the temperature of the water when the sample was taken, and other findings. The sample of water must be further treated as quickly as possible. If delays are unavoidable, each sample is preserved with a few drops of chloroform, which at least inhibits the multiplication of bacteria. But the samples should never be stored for several days because this alters the chemistry and the flasks themselves are contaminated with bacteria and algae and are then of only limited value.

Flasks contaminated in this way should be cleansed by shaking with pure sea-sand and water and then rinsed out with water and distilled water. Sample bottles which are used often should be cleaned at certain times with chromic-sulphuric acid (caution!) or with detergent and then left standing for some time with frequently renewed water.

Determination of the hydrogen ion concentration (pH) and of the free carbon dioxide (CO_2) and the oxygen content should be carried out immediately on the spot, and the remaining quantitative estimations follow in the laboratory. All the qualitative estimations can be done on the water-body.

2. Determination of the Hydrogen Ion Concentration (pH Value)

pH is the abbreviation of *potentia hydrogenii* and indicates the strength of the hydrogen ions; by "strength" is meant their concentration. The most chemically pure water is partly dissociated, and at a temperature of 22°C 10^{-14} g molecule of water per litre is equally dissociated into H^+ and OH^- ions. The concentration of the hydrogen ions thus $= 10^{-7}$. Instead of the inconvenient number 10^{-7} the pH value is expressed as the negative logarithm of the hydrogen ion concentration. pH 7 indicates neutral water, pH 7–14 alkaline, and pH below 7 acid. Naturally occurring water varies round pH 7. Extreme values are pH 12 in the alkaline range and pH 3 in the acid range.

Procedure

We content ourselves with the determination on the spot. Merck's universal indicator serves best. It provides an accuracy of 0·5 units which suffices in the first instance for hydrobiological examinations of water. The determination proceeds on the principle that certain compounds, the so-called indicators, are coloured differently by acid or alkaline water. One puts 8 ml of the water to be tested in a small white porcelain dish and adds to it 5 drops of the indicator. The mixture is carefully rotated and the colour resulting is compared with the colour scale provided with the indicator. Do not work in direct sunlight.

Other methods

To investigate special problems, e.g. the alteration of the pH within 24 hr by the photosynthesis of algae or higher plants in a lake or stream, the accuracy of the method given above is obviously not sufficient. The firm of Hellige, Freiburg in Breisgau, have developed a colour comparator which also uses a colorimetric procedure. Special indicators are used, the colouring of which in the water-sample is compared with coloured discs. They have an accuracy of 0·2 values ±0·1 unit. For still more accurate estimations in the laboratory various electrometric procedures are available, and the more robust forms of these can be used also in the open, but they do not then work with greater accuracy than the colorimetric procedures described.

3. Determination of Free Carbon Dioxide (CO_2) in Water

Rain water contains, according to its solution equilibrium with the atmospheric air and the absorption coefficient of water for carbon dioxide, about 0·6 mg carbon dioxide per litre. The carbon dioxide combines with the water partly to form carbonic acid (H_2CO_3), which is further partly dissociated again into H^+ and HCO_3^- ions. Whilst the precipitated water percolates through the layer of vegetation on the soil, additional carbon dioxide is dissolved out of the soil air. Ground and spring water are therefore especially rich in carbon dioxide which is given off at the outlet of the spring until there is equilibrium with the atmosphere. Calcium is readily soluble in water containing carbon dioxide with the production of calcium bicarbonate ($Ca(HCO_3)_2$), which only remains in solution if it is in equilibrium with a certain amount of carbon dioxide in the water, the equilibrium carbon dioxide. Only the excess of carbon dioxide over the amount of the equilibrium carbon dioxide, the aggressive carbon dioxide, makes further calcium dissolve in the water. If, on the other hand equilibrium carbon dioxide is removed from the water as was mentioned in the case of the outlet of spring water, then the bicarbonate is decomposed and weakly soluble calcium carbonate is deposited:

$$Ca(HCO_3)_2 \rightarrow CaCO_3 + H_2O + CO_2$$

Because carbon dioxide is produced by this process it continues only so long as sufficient equilibrium carbon dioxide is present in the water. Below the outlets of springs thick deposits of calcium are often laid down (Travertin) which may soon form rocks. The deposition of the calcium usually first begins some metres below the outlet of the water, after the aggressive carbon dioxide has been given off to the atmosphere.

In the following estimation of the free carbon dioxide in the water we thus measure the equilibrium and also the aggressive carbon dioxide. The richer the water is in calcium, the higher in general is its content of carbon dioxide. Because the pH value also depends on the content of free carbonic acid in the water, and because the system carbonic acid–bicarbonate in the water has a high buffer action, the carbon dioxide, the pH, and the bicarbonate content are directly related to one another.

Procedure

100 ml of water to be tested are mixed with 20 drops of 30% solution of sodium potassium tartrate and 3 drops of a 0·1% alcoholic solution of phenolphthalein. With a pipette graduated in 0·1 ml, drops of N/20 NaOH (Merck's N/10 NaOH supplied in ampoules suitably diluted with distilled water) are added and the mixture is carefully rotated every time. If free carbon dioxide is present in the water the reddish colour resulting from the addition of the sodium hydroxide disappears again. Sodium hydroxide is added until the faint reddish coloration persists in the whole test sample for 3 min, this being checked by inspection from above against a white background.

Calculation and statement of the results

The number of millilitres of N/20 NaOH required multiplied by 22 gives the content of free carbon dioxide in milligrams per litre. The results are calculated to whole numbers of milligrams and stated as free carbon dioxide.

4. Determination of the Oxygen Content of Water

The solubility of oxygen in water depends on the temperature, so that the oxygen content corresponding to the solution equilibrium varies with the temperature: more oxygen is dissolved in cold than in warm water, so that it is necessary to take the temperature when the sample of water is taken. The solubility of oxygen is also dependent on the atmospheric

Table 1. New Table for Oxygen Saturation Concentration

according to Truesdale, Downing, and Lowden (1955) for a total pressure of the atmosphere saturated with water vapour of 760 torr, in mg/l O_2

t (°C)	0·0°	0·1°	0·2°	0·3°	0·4°	0·5°	0·6°	0·7°	0·8°	0·9°
0	14·16	14·12	14·08	14·04	14·00	13·97	13·93	13·89	13·85	13·81
1	13·77	13·74	13·70	13·66	13·63	13·59	13·55	13·51	13·48	13·44
2	13·40	13·37	13·33	13·30	13·26	13·22	13·19	13·15	13·12	13·08
3	13·05	13·01	12·98	12·94	12·91	12·87	12·84	12·81	12·77	12·74
4	12·70	12·67	12·64	12·60	12·57	12·54	12·51	12·47	12·44	12·41
5	12·37	12·34	12·31	12·28	12·25	12·22	12·18	12·15	12·12	12·09
6	12·06	12·03	12·00	11·97	11·94	11·91	11·88	11·85	11·82	11·79
7	11·76	11·73	11·70	11·67	11·64	11·61	11·58	11·55	11·52	11·50
8	11·47	11·44	11·41	11·38	11·36	11·33	11·30	11·27	11·25	11·22
9	11·19	11·16	11·14	11·11	11·08	11·06	11·03	11·00	10·98	10·95
10	10·92	10·90	10·87	10·85	10·82	10·80	10·77	10·75	10·72	10·70
11	10·67	10·65	10·62	10·60	10·57	10·55	10·53	10·50	10·48	10·45
12	10·43	10·40	10·38	10·36	10·34	10·31	10·29	10·27	10·24	10·22
13	10·20	10·17	10·15	10·13	10·11	10·09	10·06	10·04	10·02	10·00
14	9·98	9·95	9·93	9·91	9·89	9·87	9·85	9·83	9·81	9·78
15	9·76	9·74	9·72	9·70	9·86	9·66	9·64	9·62	9·60	9·58
16	9·56	9·54	9·52	9·50	9·48	9·46	9·45	9·43	9·41	9·39
17	9·37	9·35	9·33	9·31	9·30	9·28	9·26	9·24	9·22	9·20
18	9·18	9·17	9·15	9·13	9·12	9·10	9·08	9·06	9·04	9·03
19	9·01	8·99	8·98	8·96	8·94	8·93	8·91	8·89	8·88	8·86
20	8·84	8·83	8·81	8·79	8·78	8·76	8·75	8·73	8·71	8·70
21	8·68	8·67	8·65	8·64	8·62	8·61	8·59	8·58	8·56	8·55
22	8·53	8·52	8·50	8·49	8·47	8·46	8·44	8·43	8·41	8·40
23	8·38	8·37	8·36	8·34	8·33	8·32	8·30	8·29	8·27	8·26
24	8·25	8·23	8·22	8·21	8·19	8·18	8·17	8·15	8·14	8·13
25	8·11	8·10	8·09	8·07	8·06	8·05	8·04	8·02	8·01	8·00
26	7·99	7·97	7·95	7·95	7·94	7·92	7·91	7·90	7·89	7·88
27	7·86	7·85	7·84	7·83	7·82	7·81	7·79	7·78	7·77	7·76
28	7·75	7·74	7·72	7·71	7·70	7·69	7·68	7·76	7·66	7·65
29	7·64	7·62	7·61	7·60	7·59	7·58	7·57	7·56	7·55	7·54
30	7·53	7·52	7·51	7·50	7·48	7·47	7·46	7·45	7·44	7·43
31	7·42	7·41	7·40	7·39	7·38	7·37	7·36	7·35	7·34	7·33
32	7·32	7·31	7·30	7·29	7·28	7·27	7·26	7·25	7·24	7·23
33	7·22	7·21	7·20	7·20	7·19	7·18	7·17	7·16	7·15	7·14
34	7·13	7·12	7·11	7·10	7·09	7·08	7·07	7·06	7·05	7·05
35	7·04	7·03	7·02	7·01	7·00	6·99	6·98	6·96	6·96	6·95
36	6·94	6·94	6·93	6·92	6·91	6·90	6·89	6·88	6·87	6·86
37	6·86	6·85	6·84	6·83	6·82	6·81	6·80	6·79	6·78	6·77
38	6·76	6·76	6·75	6·74	6·73	6·72	6·71	6·70	6·70	6·69
39	6·68	6·67	6·66	6·65	6·64	6·63	6·63	6·62	6·61	6·60
40	6·59	6·58	6·57	6·56	6·56	6·55	6·54	6·53	6·52	6·51

pressure; it is

$$S^1 = S\frac{760}{P},$$

S^1 being the saturation values corrected for the average local pressure. At higher altitudes this correction is absolutely necessary in the statement of saturation values.

In the Falkau region, at a height of about 1000 m, the correction factor is 1·13, i.e. a saturation value of 60% must be corrected to 67·8%. Tables 1 and 2 give the correction factors for various altitudes together with the oxygen-saturation values for temperatures of 0–40°C.

TABLE 2. CORRECTION FACTORS FOR OXYGEN SATURATION IN VARIOUS LAKE ALTITUDES

Metres	Pressure (mm)	Factor	Metres	Pressure (mm)	Factor
0	760	1·00	1300	647	1·17
100	750	1·01	1400	639	1·19
200	741	1·03	1500	631	1·20
300	732	1·04	1600	623	1·22
400	723	1·05	1700	615	1·24
500	714	1·06	1800	608	1·25
600	705	1·08	1900	601	1·26
700	696	1·09	2000	594	1·28
800	687	1·11	2100	587	1·30
900	679	1·12	2200	580	1·31
1000	671	1·12	2300	573	1·33
1100	663	1·15	2400	566	1·34
1200	655	1·16	2500	560	1·36

The estimation of oxygen in the water depends on the fact that sodium hydroxide together with manganous sulphate gives a white precipitate of magnanous hydroxide.

$$MnSO_4 + 2\,NaOH \rightarrow Mn(OH)_2 + Na_2SO_4$$

If oxygen is present at the same time, the manganous hydroxide is oxidized to brown-coloured manganese oxyhydrate ($MnO(OH)_2$), and this occurs to a degree proportional to the amount of oxygen. The approximate content of oxygen can be judged from the brown coloration. In accurate titration procedures, potassium iodide–sodium hydroxide is added. In strongly acid media (with a pH below 1), produced by adding sodium bisulphate, manganic ions are freed and react with the iodine ions from the potassium iodide to form free iodine:

$$MnO(OH)_2 + 4\,NaHSO_4 + 2\,KI \rightarrow I_2 + MnSO_4 + K_2SO_4 + 2\,Na_2SO_4 + 3\,H_2O$$

The amount of free iodine is equivalent to the amount of oxygen present in the sample. By titrations with sodium thiosulphate the amount of iodine can be determined and the oxygen content can be calculated from this.

Reagents

1. Solution of manganous sulphate: 100 g of $MnSO_4 \cdot 4\,H_2O$ plus 200 ml of boiled distilled water.

2. Potassium iodide–sodium hydroxide: 100 g of NaOH+200 ml of boiled distilled water+50 g of potassium iodide.
3. Sodium bisulphate solution: 500 g of $NaHSO_4 \cdot H_2O$+400 ml of boiled distilled water.
4. Syrupy phosphoric acid: 89%.
5. Sodium thiosulphate solution: a 0·01 N solution, always prepared before use from 0·1 N solution (Merck's ampoules).
6. Starch solution: mix 1 g of soluble starch with water to a fluid paste, transfer to 100 ml of hot water and, after cooling, add 0·5 ml of 40% formalin.

Procedure

An "oxygen bottle" of about 100 ml capacity and of known volume is filled with the water to be tested, free from air bubbles. To do this one uses a glass funnel with a rubber tube attached which reaches down to the bottom of the bottle. The water to be tested is added through the funnel in such a way that the cylinder is flushed through several times. The rubber tube is slowly withdrawn and the glass stopper, the end of which has been obliquely ground down (Fig. 15), is put in in such a way that no air bubbles remain in the bottle. The bottle is then carefully opened again and to it 0·5 ml each of manganous sulphate and potassium iodide–sodium hydroxide solution are added, the end of the pipette being introduced to 1 cm under the meniscus of the fluid in the bottle. The bottle is then finally closed, mixed by shaking, and a wire clip (Fig. 15) is put on it to keep the stopper in place during transport. The bottles thus treated should be transported as much as possible in damp

FIG. 15. Oxygen bottles with bevelled-off glass stoppers and closure clip.

cloths and in the dark. In the laboratory the bottles are opened after the precipitate has settled down and 3 ml of sodium bisulphate solution are put in. The bottles are closed again and shaken to dissolve the precipitate. After 10 min the contents of the bottle are transferred quantitatively to a 200-ml Erlenmeyer flask. Add 1 ml of starch solution to the sample and titrate with N/100 thiosulphate solution until it is decolorized. In water rich in iron, 3 ml of syrupy phosphoric acid are used to dissolve the precipitate instead of the sodium bisulphate solution.

Calculation and statement of the results

The oxygen content per litre is computed by the formula

$$\text{mg}\,O_2/\text{l} = \frac{nF80}{V-v}$$

in which n is the number of ml of thiosulphate needed, F is the titration factor of the thiosulphate solution (= about 1), V is the exact volume of the oxygen flask used, and v is the total volume of the $MnSO_4$ and NaOH added, i.e. 1. From the oxygen content thus calculated, 0·07 mg is subtracted for the oxygen introduced with the reagents. The satura-

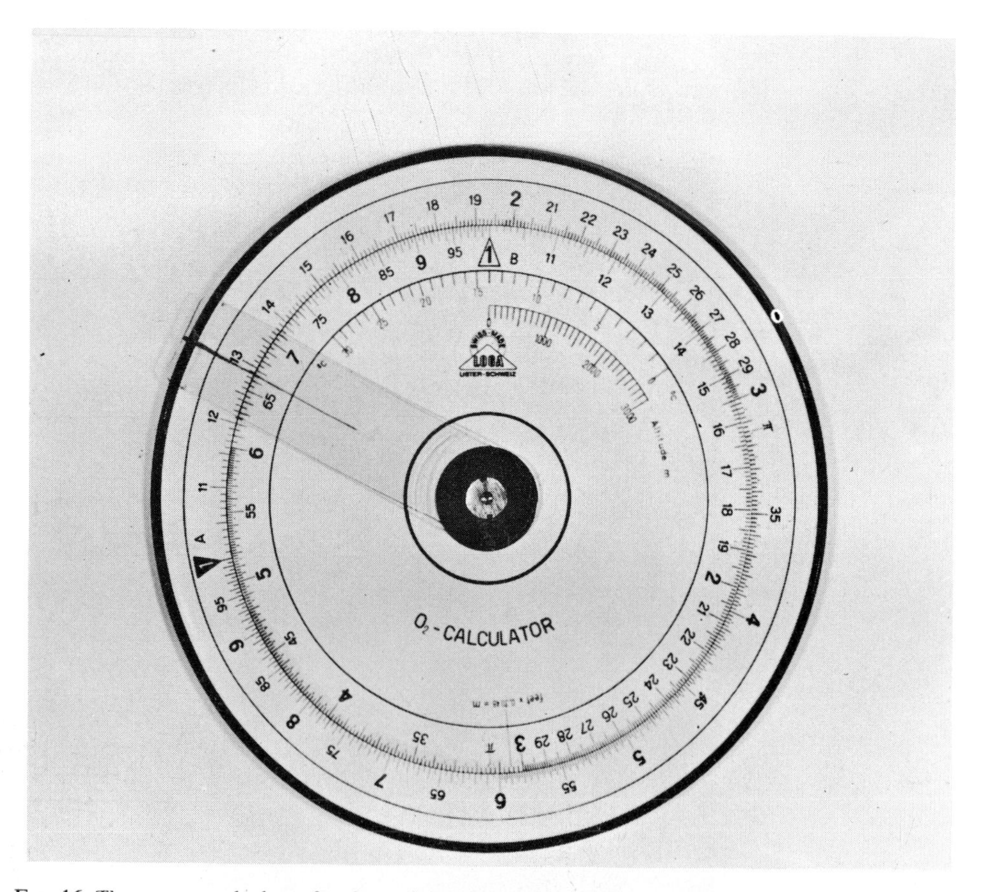

FIG. 16. The oxygen calculator for determining the oxygen content of water and the saturation value.

tion points are to be calculated according to the milligram values obtained and to the theoretical value (= 100% for the temperature of sampling, for which, if necessary, a correction for altitude must be borne in mind (see Tables 1 and 2).

Results are stated in mg/l to one decimal place or are expressed in percentage saturation of the whole. This calculation is easily done with the oxygen calculator on the slide-rule principle (Fig. 16). If the oxygen content is to be given in ccm/l, the milligram value must be multiplied by 0·7 and conversely 1·43 times. The Winkler method works with an error of 3–5% (Carpenter, 1965). An improved technique which reduces the error to 0·1% is described by Carpenter, but it is not suitable for use in the field.

Preliminary estimation in the field

The techniques of Winkler and Ohle given above can be carried out with sufficient accuracy only in the laboratory. It is possible with the following method to make an estimate, though a rough one, of the oxygen conditions at the water-body. An oxygen bottle is filled with 100 ml of the water sample and 2 ml each of manganese sulphate and sodium hydroxide are added. The bottle is then closed and shaken. The resultant precipitate is coloured more or less brown according to the oxygen content of the water. If the precipitate remains white, no oxygen—or only a very little—is present. A light yellow precipitate shows that the sample is poor in oxygen, a brown one that it is rich in it. There is a colour chart by Plehn, which is illustrated by Liebmann (1962).

Other methods

For accurate limnological investigations the iodine difference method of Ohle is sometimes used. The samples must be, however, titrated after 10 min, a requirement that can, in field operations, only be met if a laboratory truck is available. In addition to titration methods a number of electrometric methods of estimation are nowadays available (see Tödt and Petsch, 1955; Ambühl, 1960; Mackereth, 1964; and McLeod, Bobblis, and Yentsch, 1965).

5. Determination of Ammonium in Water

Inorganic nitrogenous compounds mostly occur in water as ammonium (NH_3), nitrite (NO_2), and nitrate (NO_3). Ammonium compounds are set free in water by the decomposition of protein and they undergo a more or less rapid bacterial oxidation to nitrite and then to nitrate. Spring and ground water not contaminated by human activities as a rule contain only nitrate and no ammonium. If this is present in readily detectable quantities (more than 0·1 mg/l), this indicates putrefactive processes in water due, for example, to still incomplete oxidation of ammonium formed somewhere or introduced. Ground water containing iron may, on the other hand, contain, under natural conditions, 3–4 mg/l of ammonium.

The method of determination depends on the fact that the ammonium ion, NH_4^+, gives a yellowish-brown colour with Nessler's reagent.

Reagents

1. A 50% solution of sodium potassium tartrate. 50 g of crystal of this are dissolved in 100 ml of distilled water. The solution is left to stand for some days and is neutralized against phenolphthalein. It should be filtered through glass wool and kept in brown glass bottles.

2. Nessler's reagent obtained ready for use from Messrs. Merck or some other firm.
3. Ammonium standard solution. Dissolve 2·966 g of NH_4Cl in 1 l. of distilled water. Each millilitre of this solution diluted 1:10 contains 0·1 mg of NH_4.

Determinations in the field (qualitative)

To 20 ml of the water-sample add 10 drops of the sodium potassium tartrate solution plus 2–3 drops of Nessler's reagent and shake vigorously. If no clearly detectable yellow coloration appears, less than 0·1 mg/l of ammonium is present; when more than this is present a more or less yellow coloration is seen (0·1–5 mg/l). A yellowish to reddish-brown precipitate forms with 5 mg/l and more present. Care is necessary with waters, such as moor or marsh water, that are themselves already coloured.

Quantitative determination

Put 100 ml of the water-sample into an Erlenmeyer flask and add 2 ml each of the sodium potassium tartrate solution and Nessler's reagent. Shake vigorously and leave for 5 min. Fill a second flask with 100 ml of distilled water (free from ammonia) and mix it with the same reagents. Then with a burette graduated in 0·1 ml, add drop by drop the standard ammonia solution and compare the resulting yellowish colour with the colour of the water-sample. When the colours match the amount of standard ammonia solution added can be read off and the ammonium content can be calculated.

Here reference may be made to the use of the Hehner cylinders. This is a specially made standing cylinder containing 105 ml and graduated in 1 ml. A lateral stopcock permits convenient adjustment of the height of the meniscus. The cylinder must be made of colourless glass because it is used for the comparison of coloured solutions. The left-hand cylinder is filled with 100 ml of the water being tested and the necessary reagents are added to it, and it is filled up to 105 ml with distilled water. Into the right-hand cylinder 80–90 ml of distilled water are put and then the standard ammonia solutions and the reagents, and this cylinder is also filled up to 105 ml. The resultant colorations are compared with each other in clear diffuse daylight at a distance of some centimetres against a white background. If the water being tested is more highly coloured than the solution it is compared with, it is drained off until the colours agree. This may happen at 60 ml. The solution for comparison contains 1 mg ammonium per litre; thus 0·1 mg per 100 ml. Then $60:0{\cdot}1 = 100:x$ and $x = 0{\cdot}17$ mg. The sample thus contains 1·7 mg of ammonium per litre. If the solution used for comparison must be drained off to 60 ml, it contains 0·06 mg. Because the water being tested corresponds in colour at this value, 100 ml of it contain 0·06 mg of ammonium and a litre of it contains 0·6 mg.

This use of the Hehner cylinder serves for all colorimetric estimation techniques. The weakest point of the method is the subjective colour comparison which naturally leads to subjective errors.

Other methods

Simpler than working with standard solutions is the use of the Hellige colour comparator which includes colour discs for the different standards. In the laboratory nowadays the coloured solutions are photometrically measured in, for example, the Elko instrument. The ammonium value for the extinctions observed are taken from a standard curve previously obtained with standard solutions. These photometric measurements are naturally more accurate and especially are free from subjective errors. Because the instruments are not available everywhere, they will not be described in detail here.

Statement of the results

The values are given to an accuracy of one decimal point in mg/l.

Calculations

mg/l NH_4 (ammonium) = mg/l NH_3 times 1·059
mg/l NH_3 (ammonia) = mg/l NH_4 times 0·944
mg/l (ammonia-nitrogen) = mg/l NH_3 times 0·823.

6. Determination of Water-soluble Phosphate

Principle. *Molybdic acid reacts on phosphation to form yellow phosphomolybdic acid which is reduced by reducing agents to a blue compound, phosphomolybdenum blue.*

Reagents

1. Ammonium molybdate–sulphuric acid. Mix every time before use 1 vol. of 10% ammonium molybdate solution with 3 vol. of 50% H_2SO_4.
2. Solution of stannous chloride. Before use dissolve (with care) 0·27 g of tin foil in 7·5 ml of concentrated fuming hydrochloric acid and fill up with distilled water to 50 ml. Only the clear supernatant solution may be used.

Procedure

To 50 ml of the water being tested, add 0·5 ml of the ammonium molybdate–sulphuric acid and 0·15 ml of the stannous chloride solution. In the presence of PO_4–P a blue colour develops, and this can be either estimated photometrically or the Hehner cylinder can be used as for the estimation of ammonium.

Preparation of the standard solutions and dilutions

1. Standard solution of phosphate. Dissolve 0·57466 g of disodium hydrogen phosphate according to Sörensen ($Na_2HPO_4 \cdot 2H_2O$) in distilled water up to 100 ml. The solution contains 1 g of PO_4–P per litre.
2. Dilution A. 1 ml of the standard phosphate solution up to 1000 ml of distilled water. 1 l. contains 1 mg of PO_4–P.
3. Dilution B. To 10 ml of the dilution A add distilled water to 100 ml. This solution contains 0·1 mg of PO_4–P per litre.
4. Comparative series:

	2	5	10	20	50	70	100 μg PO_4-P/litre
Dilution B	1	2·5	5	10	25	35	50 ml
+distilled water	49	47·5	45	40	25	15	0 ml

7. Determination of the Alkalinity (SBV) and the Hardness due to Carbonate (Combined Carbon Dioxide)

(Cf. the procedures for carbon dioxide pp. 22 ff.)

The carbonate content of the water is determined by titration of 100 ml of the water with a tenth-normal solution of hydrochloric acid against methyl orange as an indicator. The carbonate occurs mostly as the bicarbonate $(HGO_3)_2$.

Reagents

1. 0·1% aqueous solution of methyl orange.
2. N/10 or N/100 HCl. Normal solutions can be obtained from Messrs. Merck under the name Titrisol.

Procedure

Put exactly 100 ml of the water being tested into an Erlenmeyer flask with 3 drops of the methyl orange solution. Titrate with N/100 or N/10 hydrochloric acid carefully until there is a colour change from yellow to orange. For comparison use a sample of the water being tested which has not been titrated, but to which methyl orange has also been added. If necessary make a duplicate determination with the first titration as a comparison.

Statement of the results and calculations

The ability to combine with acids is stated as the number of millilitres of N/10 HCl required for 100 ml of the water tested.

Further: The amount of N/10 HCl in ml multiplied by 2·8 = the carbon hardness. The amount of N/10 HCl in ml multiplied by 22 = the firmly combined carbonic acid.

8. Determination of the Hardness due to Calcium

Literature: Schwarzenbach, G. and Ackermann, H., *Helv. Chim. Acta* **31,** 1029 (1948). Gad, G. and Fürstenau, *Gesundheitsingenieur* **74,** 191 (1953). *Deutsche Einheits Verfahren,* edited by L.-W. Haase, Weinheim, 1954.

Calcium and magnesium ions form internal anion complexes with the disodium salt of ethylene-diaminetetracetic acid (EDTA). The end point of the reaction is shown by a change in the colour of suitable indicators.

Reagents

1. N/10 HCl in Merck's ampoules.
2. 30% NaOH.
3. Murexide indicator, dilute aqueous solution, freshly prepared.
4. EDTA solution. Dissolve 6·65 g of the disodium salt.

Procedure

The sample of water used to determine the alkalinity can be used. Add 0·5 ml of N/10 HCl to the sample to be titrated and boil off the carbonic acid formed for 10 min. Then to the hot sample, about 50°C, add 0·2 ml 30% NaOH and 1 ml of the murexide solution. Then titrate with the EDTA solution until the colour changes from red to bluish violet.

Calculation and statement of the results

The amount of EDTA solution needed gives directly the calcium hardness in German degrees of hardness (dH°). 1 ml of EDTA solution = 1 mg of CaO/100 ml of the water tested = 10 mg of CaO/l = 1 dH°. Care should therefore be taken to use exactly 100 ml of the water tested. The results are given rounded off to whole milligrams.

9. Rapid Method for the Estimation of the Total Hardness of Water

The method here given can safely be carried out on excursions into the country if flasks with a burette attached for titration are used.

Reagents

1. Titriplex solution A (Merck) for hard water of more than 3 degrees of hardness. Titriplex solution B for soft water up to 3 degrees of hardness.
2. Indicator tablets made by Merck.
3. Take along concentrated ammonia solution ($D = 0{\cdot}910$) in pipette bottles and estimate beforehand the number of drops per millilitre.

Procedure

A. For waters in areas where there is limestone (shell limestone, Red marl, Jura, chalk, etc.) put 100 ml of the water to be tested into an Erlenmeyer flask and dissolve an indicator tablet in it. Add 1 ml of the ammonia solution and titrate with Titriplex solution A until the red colour changes to green.

B. In waters poor in lime (calcium) where there is primitive rock, mottled sandstone, etc., test as under A but before titration heat to 40°C. According to Höll one should add a few drops of a solution of sodium phosphate. Titrate with Titriplex B until the colour changes to green.

Recently Messrs. Heyl of Hildesheim issued ready-made tablets for the determination of the hardness of water under the names "Durognost T" for the total hardness and "Durognost C" for the carbonate hardness. To 100 ml of the water to be tested, add an indicator tablet and as many of the titration tablets as are required to change the colour of the indicator. Titration tablets are issued in a strength of 5 dH° for hard water and 1 dH° for soft water. If, for example, one 5° tablet and three 1° tablets are required, the hardness of the water is 8 dH°.

Calculation and statement of the results

For the procedure A the total hardness is calculated from the number of millilitres. Titriplex solution A used multiplied by 5·6. According to Höll the results are about 5% too low. For procedure B the number of millilitres of Titriplex solution B required directly gives the total hardness. The results are given in German degrees of hardness accurate to 0·1 dH°.

10. Estimation of the Potassium Permanganate Consumption (Approximate Method)

The method permits a rough estimate of the organic substances in the water which are especially important in contaminated waters. It depends on the fact that the oxygen set free from potassium permanganate by the addition of sulphuric acid oxidizes the organic material

$$2\,KMnO_4 + 3\,H_2SO_4 \rightarrow 2\,MnSO_4 + K_2SO_4 + 3\,H_2O + 5\,O$$

The potassium permanganate is thus reduced. The potassium permanganate consumption needed for the oxidation of the organic substances can be found by the disappearance of the violet colour. Höll gives the following simple method which can be carried out in excursions with the simplest apparatus, and gives sufficiently accurate results for orientation.

Reagents

1. Dilute sulphuric acid: 1 part of the concentrated acid and 3 parts of distilled water.
2. N/100 solution of potassium permanganate: dilute a N/10 solution (Titrisol made by Merck) with distilled water to make it.

Procedure

Put 10 ml of the water to be tested into a test tube and add 5 drops of the dilute sulphuric acid and 3 drops of the N/100 solution of potassium permanganate. After shaking the mixture, let it stand for about 5 min and then heat it carefully to boiling. In order to avoid the boiling liquid spattering out, the test-tube must be constantly kept in vigorous motion. Water with permanganate values above 30 mg/l shows at ordinary temperature (without boiling) a decolorization in a few minutes. At values of more than 50 mg/l, 5–6 drops of the permanganate solution are also decolorized without heating. Waters with 20–30 mg/l permanganate consumption are decolorized only after boiling. Water with permanganate values below 12 mg/l are not even decolorized when they are boiled and allowed to stand longer.

CHAPTER 2

METHODS FOR THE INVESTIGATION OF THE OPEN WATER ZONE OF STANDING WATERS (PELAGIAL)

A. METHODS FOR THE INVESTIGATION OF PLANKTON

V. Hensen (1895) defined plankton as all those living organisms which float "willy-nilly" in free water and are independent of the shore and the bottom. The first knowledge of this living community was obtained when Johannes Müller for the first time pulled seawater through a net made of fine-meshed silk gauze, of the kind used in mills to sift the flour, and this collected an enormous number of unknown organisms of quite remarkable structure. Apstein (1896) and Zacharias (1907) then used this method with similar results for the investigation of inland waters. Very soon plankton science became popular and a fashionable branch of research.

In recent years many suggestions have been made for the classification of the plankton into size classes. Margalef (1955) and Peres and Deveze (1963) give the following subdivisions for freshwater and marine plankton:

	Margalef (freshwater)	Peres and Deveze (marine)
Ultraplankton	below 5 μ	below 5 μ
Nannoplankton†	5–50 μ	5–60 μ
Microplankton	5–500 μ	60–1,000 μ
Mesoplankton	0·5–1 mm	1–5 mm
Macroplankton	above 1 mm	above 5 mm
Megaplankton		very large forms

Dussart (1965) suggested the following uniform classification for limnological and marine plankton:

Ultra nannoplankton	below 2 μ	Nannoplankton in the wider sense
Nannoplankton *sensu stricto*	2–20 μ	
Microplankton	20–200 μ	Net-plankton or filtrable plankton
Mesoplankton	200–2000 μ	
Megaplankton	above 2000 μ	

† Fraser (1965) has rightly pointed out that it should more correctly be nanoplankton, from the Greek *nanos*, small, minute; nevertheless, in hydrobiology the incorrect term of Lohmann is universally used.

Neither the plant nor the animal plankton is distributed uniformly in the water. The plant plankton (phytoplankton) which consists essentially of algae, remains in the epilimnion in so far as there is sufficient light for assimilation. The animal plankton (zooplankton) is less restricted to the epilimnion and is found at all depths of the water. Besides, most zooplankton species have powers of independent movements, although these are feeble, so that they can alter the depths at which they live. Many of them, especially copepods and species of Cladocera, do this periodically and perform regular daily vertical migrations in that they live during the day at a certain depth and towards evening rise to the surface and at midnight sink again a little and towards morning seek again their daytime depth. There is an abundant literature about this behaviour, which is in the first instance determined by the light. This literature is detailed by Ruttner (1962), Schröder, Siebeck, and Ringelberg.

Horizontal variations in the distribution of plankton in a lake arise from active migration of the plankton animals out of the shore zone, the flight from the shore, which was first correctly explained by Siebeck (1964); and further by a classification of plankton horizons into individual clouds of plankton due to local water currents (Schröder, 1962) and also temporally displaced by developmental rhythms of the plankton organisms in different parts of a lake.

Theoretical knowledge of this uneven distribution of the plankton is extraordinarily important for practical work. In all accurate qualitative and quantitative investigations of the distribution of the plankton, these facts must be taken into consideration; otherwise there is a risk of concluding from a single sample of plankton that there is a distribution of the plankton that does not in fact exist.

The methodological development of the plankton studies is concerned chiefly with two problems—the quantitative taking of representative samples of the plankton from the water and the enumeration of the organisms contained in the samples. By quantitative we mean that the instruments put into the water being examined completely capture the plankton in it; by representative we mean that the amount and composition of the plankton captured are suitable for the planktonological characterization of the water and are ensured by putting in suitable instruments so that, for example, in the vertical and horizontal distribution picture a maximum or minimum is not taken on which statements are later made. The enumeration of the organisms captured seems above all to be trivial, but the methodology of it is very much differentiated and is, for phytoplankton especially, always being improved.

The practice of taking samples and further treatment of these is so different for plants and animals that the two groups will here be treated separately. For zooplankton, methods for direct observation and recording in lakes have been recently introduced.

ZOOPLANKTON

I. DIRECT OBSERVATION AND RECORDING OF ZOOPLANKTON IN THE WATER

The simplest method of determining the distribution of zooplankton in water is direct observation from the boat or by diving. There is a widespread prejudice against a simple method that it is primitive and therefore must fail to give noteworthy results; but Schröder has reported observations made in this way, and he was the first to observe directly the

FIG. 17. Exposure section of the underwater television camera with the objective and the two light projectors. (Photo by Schröder.)

uneven distribution of the zooplankton and the individual clouds of plankton in Lake Constance and the lakes of the Black Forest.

A further possible method of directly recording the distribution of the plankton in lakes is by using the underwater television camera and the echo depth-sounder.

Elster and von Brandt tested the introduction of the television camera for limnology in the Königsee, and Schröder was the first to make direct plankton observations with its help. For directions for the use of the television camera (Fig. 17), the works of these authors, and also Hunger (1957) and Ohle (1959), must be consulted.

The echo-sounder (Fig. 18) has been used for a long time in fisheries. The instrument works on the following principles: a transmitter sends out supersonic impulses which are partly reflected by the bottom and by "obstacles" in the water (e.g. any fish) and are again taken up by the instrument which continuously records them. Horizons and clouds of plankton can be recorded (Figs. 19 and 20) when certain requirements are fulfilled. In this book we cannot enter into the complex methodology. Schärfe (1952) and Shiraishi (1960) were the first to demonstrate this possibility, and it was further developed by Schröder (1961) who used it with good results for accurate study of the distribution of plankton in Lake Constance.

Naturally it is not possible to determine by these direct observations species of the plankton, but such methods are especially suitable for recording temporary variations in the distribution of the plankton and for checking the picture of the various layers obtained by other methods (cf. Fig. 19).

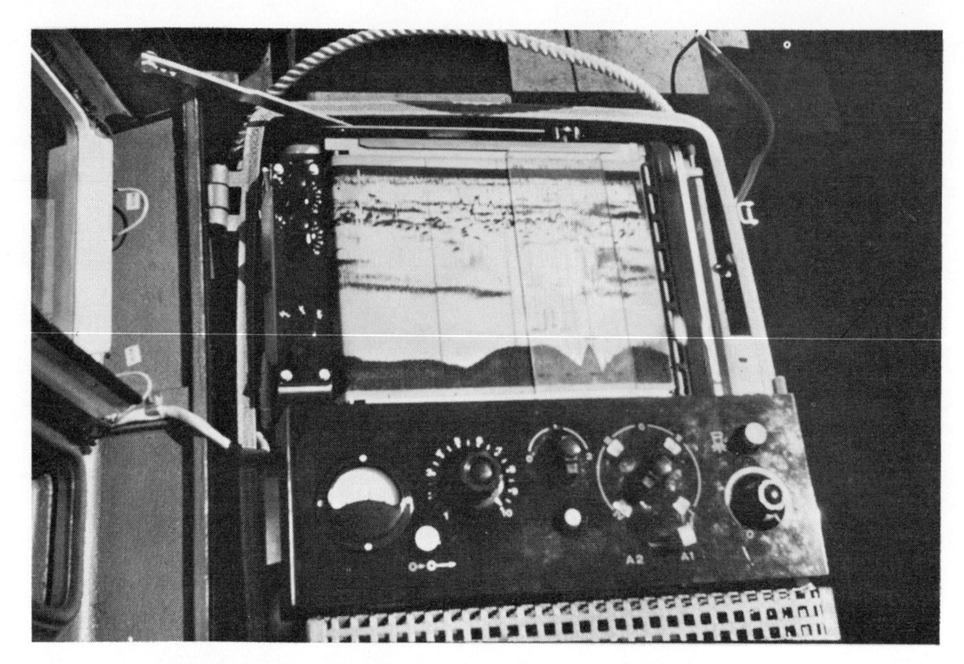

FIG. 18. Echo-sounder. (Photo by Schröder.)

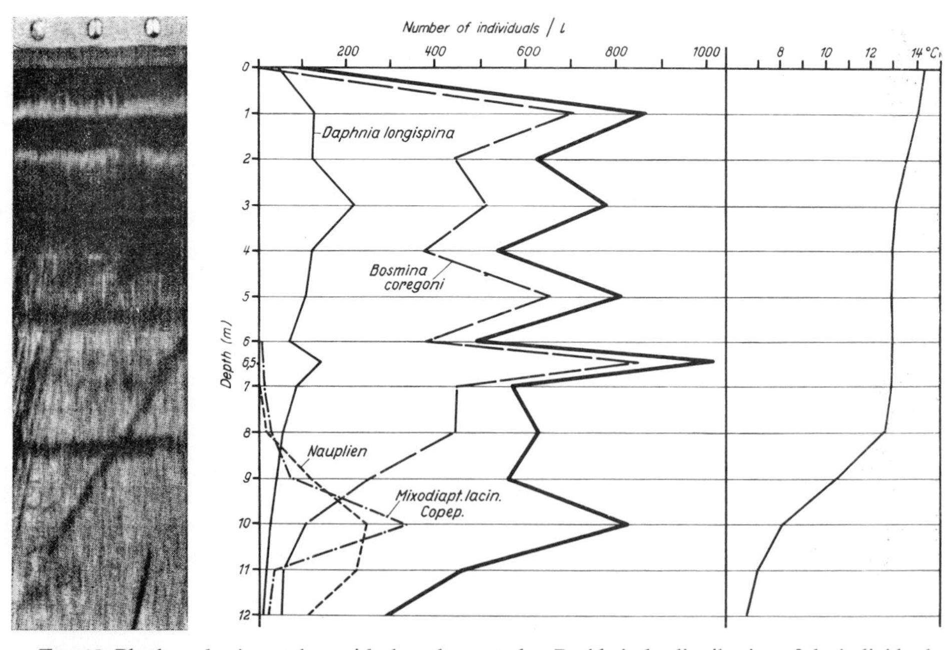

FIG. 19. Plankton horizon taken with the echo-sounder. Beside it the distribution of the individual species checked by means of catches made with the pump. On the extreme right the temperature stratification. Abscissa: number of individuals per litre; ordinate: depth. (Original by Schröder.)

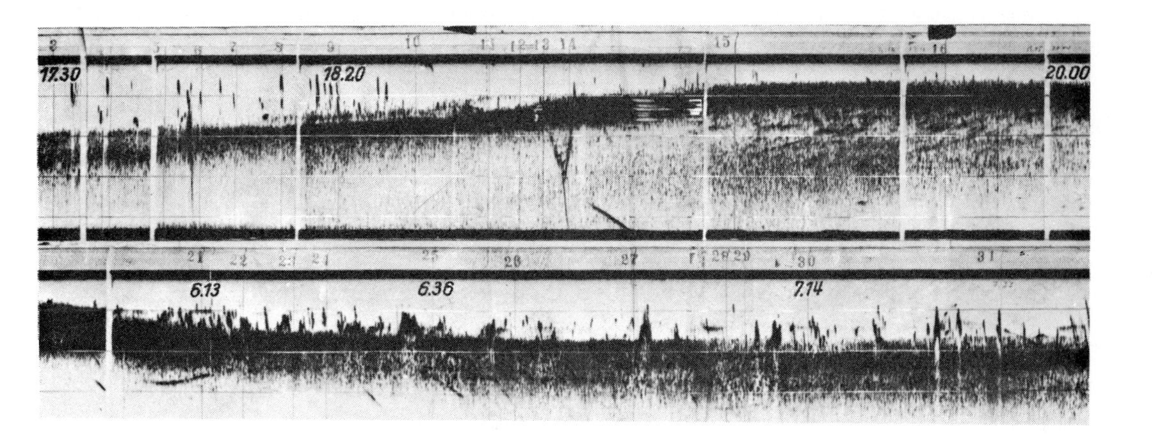

FIG. 20. Echogram of the vertical migrations of the larvae of the midge (*Chaoborus*). From above left to below right, the animals ascend together and then migrate again down to the depths; cf. the times specified. (Original by Schröder.)

II. COLLECTION OF ZOOPLANKTON FROM WATER

Basically three methods of doing this are available:

1. Collection with nets.
2. Collecting with water-bottles.
3. Collecting with water-pump.

With nets the plankton is filtered out of the water, and by the other two methods the water is collected together with the plankton. These latter methods therefore have the advantage that the relationship between the amount of plankton and the volume of water is given directly; but here lies the greatest difficulty in the use of nets, and to obviate it the incorporation of counting devices in the net has been tried.

1. COLLECTION WITH PLANKTON NETS

The net acts as a filter and is, like any other filter, liable to become progressively choked up. This depends on the following factors amongst others: the ratio of the opening of the net to the filtering surface, the fineness of the filter, and the density of the plankton in the water. The coarser the net is the less likely it is to become choked up, but it will then trap only the larger organisms and all other will pass through it. The result is that with a wide-meshed net one captures chiefly plankton animals but not the larvae of these nor other small plankton (rotifers, unicellular organisms, phytoplankton). Plankton nets of this kind can be so constructed that the ratio between the opening of the net and the filtering surface is as 1 : 4 (for the manufacture, see below). For collecting smaller organisms, nets with smaller meshes must be used and the filtering surface must be increased in relation to the net opening in such a way that the very long zeppelin net arises (cf. phytoplankton and Fig. 40).

Nowadays bolting cloths are available made from silk, perlon, and nylon in various grades and fineness. These are offered with statements of the diameters of the pores or the number of meshes per square centimetre. Table 3 gives the most important varieties. In the table note should be taken of the difference in mesh width.

TABLE 3. COMPARISON OF THE INSIDE WIDTHS OF THE MESHES OF DIFFERENT KINDS OF BOLTING CLOTH (in μ)

No.	Silk gauze	Monodur perlon	Monofilament nylon	Open sifting area (%)
0	490	500	500	54
3	285	280	300	50–54
8	195	200	180	44–47
12	106	112	112	34
16	74	71	87	22
20	63	63	75	20
25	55	56	65	20
20 GG	1000	1000	—	66
30 GG	670	670	—	62
53 GG	328	335	—	45

The last column gives the sifting area in per cent of the total surface of a piece of gauze. By "sifting area" we mean the sum of the mesh area per unit area of gauze. It can be seen from the sifting table that the area decreases with the fineness of the gauze. This causes one of the difficulties of quantitative collection with nets.

Nowadays nylon or perlon nets are used, but silk nets are no longer in use; their fibres swell up in water and narrow down the meshes; with the artificial fibres, which are not twisted, this does not happen (Figs. 21a, 21b).

Nets are very good for qualitative studies of the plankton of a lake. For quantitative studies, which must be accurate if they are ever done, the use of nets has recently been considerably improved. Nevertheless, one should be nowadays fully aware that the distribution of plankton in lakes is not uniform either horizontally or vertically, and should completely abandon nets for quantitative studies in scientific research and work instead with the Clarke–Bumpus sampler, the water-bottle, or the pump. Fleminger and Clutter (1965) have investigated whether zooplankton flees before the opening of a net moved along or turns aside from it—a problem of the greatest importance for plankton methods. In an artificial pool with accurately known plankton density and distribution, small nets were more markedly avoided, but the varying species behaved differently. The time of day had no effect on the flight behaviour, so that it is not optically determined.

Because often no apparatus other than the plankton net is available, we may here—in spite of the considerations brought forward—discuss accurate net-methodology. Exploratory quantitative samples can at any rate be taken with the net, especially with the closing net, which is only drawn vertically through short distances. These "quantitative" samples should not, however, form the basis of scientific statements.

Plankton nets of many sizes and many widths of mesh are commercially available, but they are very costly. Very servicable, especially on excursions, are the small Kosmos plankton nets, made in the degrees of fineness Nos. 8, 12, 16, 20, and 25. For more accurate studies of the zooplankton the best is a so-called "normal net" with a filtration surface of 1 m^2 and a net opening of 0·25 m^2.

Making a normal net according to Wagler

A large piece of No. 8 gauze measuring 1 m^2 is cut out as shown in Fig. 22. The piece comes as 102 cm broad from the makers. The two smaller marginal pieces *b* and *c* are sewn

(b)

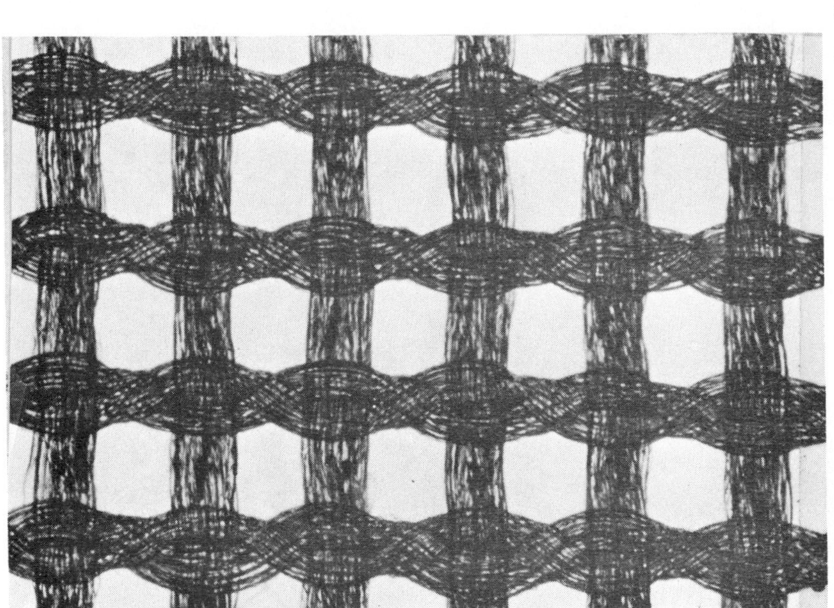

(a)

FIG. 21. Net gauzes of the same mesh width: (a) from silk; (b) from nylon. The twisted fibres of the silk gauze swell up with use.

together so that a piece like the section *a* is formed. Each of the two pieces is given an upper band made of strong linen about 15–20 cm broad. Both parts are now sewn together to form the bag of the net; the seam must be but in with special care. The inside of the bag of the net must be made absolutely smooth, so that no plankton can adhere to it. It is best

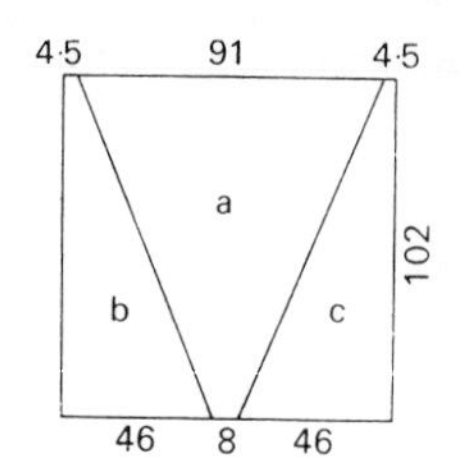

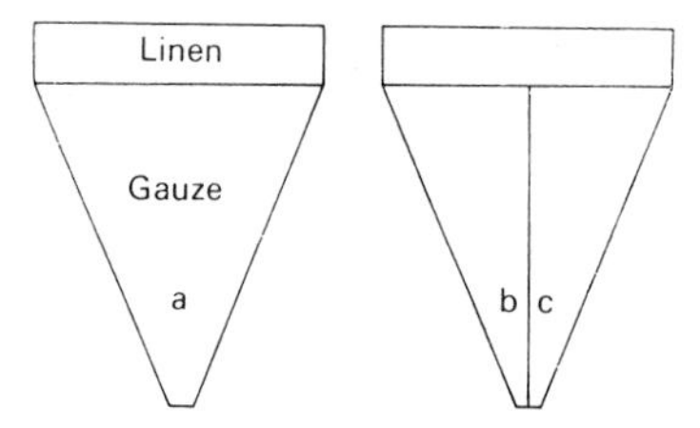

FIG. 22. The cuts for making a normal net. (After Wagler.)

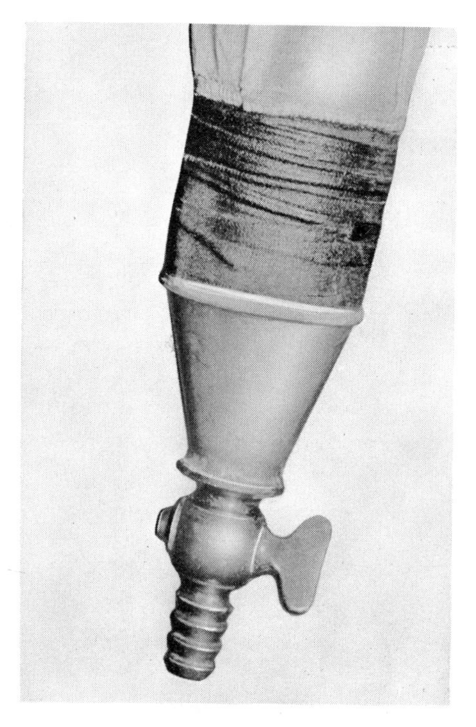

FIG. 23. The net funnel with its stop-cock for taking net samples.

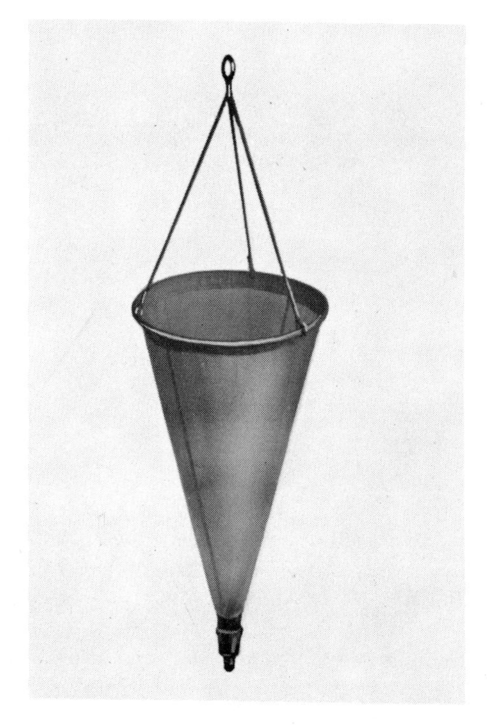

FIG. 24. Normal net. (After Wagler.)

therefore to work with right to left seams put in on the outside. The net ring, 56·6 cm in diameter, is sewn into the upper linen border, and the funnel is tied into the lower opening of the net, i.e. tied round with strong twine (Fig. 23). It has a diameter of 35 mm and a vertical measurement of 60–70 mm. To fix it better to the net it has a small projecting ridge on its upper border. Below is an outlet nozzle provided with a stopcock.

The net is now provided with three suspension lines, tied to the net ring at exactly equal distances from each other and attached to a perlon line (Fig. 24). It is advisable to attach the suspension lines to a small brass ring to which the draw line is also tied or, better, a swivel is used. In this way several nets can if necessary quickly be attached to the line. For many purposes it is useful to use a draw line marked at intervals of a metre. The marking is done with multicoloured threads, as described on p. 14. The work involved is worth while because the line can also be used for other work in the lake, and indicates at any time the depth at which instruments sent down operate.

For nets made of finer gauze the relation between the net opening and the filtration surface must be altered in favour of the filter surface; with the same net opening the net must be longer the finer the gauze is. In practice a narrower net opening than that of the normal net is selected. The zeppelin net used in the Falkau Institute has a ring diameter of 15 cm and a length of 1 m and is made with No. 25 gauze (see Fig. 40).

Steuer (*Planktonkunde*, 1910) gives the following mathematical directions for the construction of any plankton net: the shape of the bag of the net is usually conical. When making the bag one must, according to Apstein's instructions, first make a model, a paper pattern. If I make the truncated cone (Fig. 25a) and mark off the height of the tip cut off with an x, then $x:x+i = r:R$, and it follows that $x = \frac{ri}{R-r}$. If we imagine the sheath of the cone rolled up (Fig. 25b), the circumference of the circle which I can draw with the radius $x(=2x\pi)$ is related to $2r\pi$ as 360°: a; so that $\frac{2x}{2r} = \frac{360}{a}$, and from this if follows that $a = 360r/x$. So far Steuer.

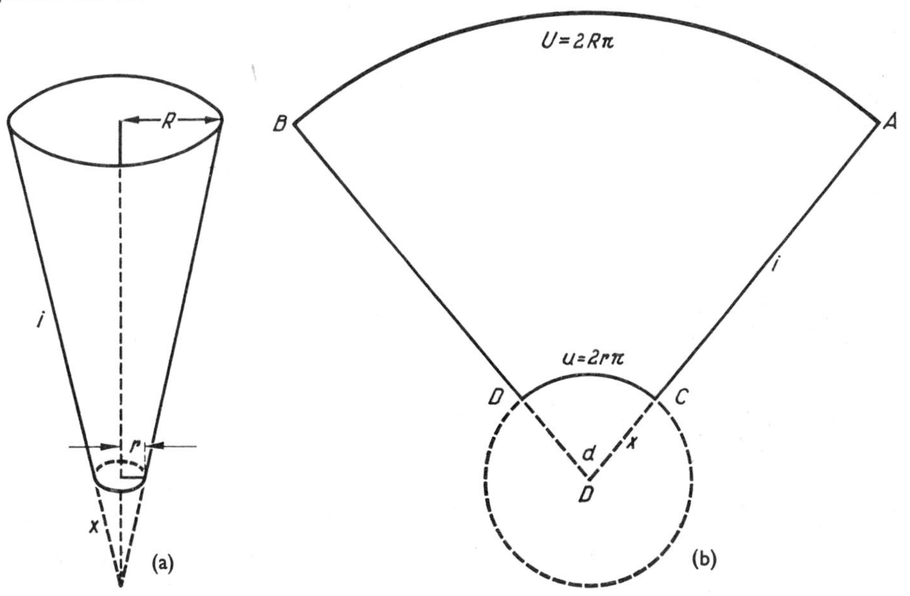

FIG. 25. The construction and cuts for making any plankton net desired. (After Steuer.)

Determination of the net factor

Because the nets undergo increasing obstruction during a catch, they do not make quantitative catches and filter less water than can theoretically flow into the net. This obstruction of the net depends on the composition of the plankton and the duration of the catch. Basically the effect cannot be calculated nor can its extent be foreseen. This is the most serious defect of quantitative net methods. A defective method, the errors of which can be exactly formulated, is quite useful. The net does not fulfil this requirement because there is no control of the errors.

If nothing but a net is available for quantitative work, it is helpful to standardize it before beginning the investigations against a method considered to be quantitatively correct. The pump method is judged to be "quantitatively correct". By it a definite quantity of water is pumped out of the lake and the plankton in it is filtered out through a gauze net (for the exact method of work see p. 47). At the same time as the sample is taken with the pump, a corresponding volume of water is fished over at a nearby similar place with the net to be tested and the results are compared with those obtained with the pump. The relation between the samples caught by the pump and net then give the net factor, which, strictly speaking, holds only for the plankton conditions prevailing during the calibration.

In order, for example, to compare the quantity of plankton in 300 l. of lake water pumped out with the collecting power of a zeppelin net with an aperture of 200 cm^2, one must thus apply a net-traction of

$$\frac{300{,}000\ cm^3}{200\ cm^2} = 1500\ cm = 15\ m\ long$$

and compare the amounts of plankton with one another.

It has often been found convenient to put a linen attachment shaped like a reversed funnel in front of the net. Behind it in the net a minimum pressure is created, so that through the inlet opening more than the calculated column of water can flow in. In comparative catches a normal net provided with such an attachment takes in, on average, twice as much as it does without it (Elster, 1958). Still it is in no sense certain that the greater yield corresponds to the correct relations. Elster (1958) published a critical survey of net catches.

The use of nets

Nets are especially excellent for use when attempts are being made to carry out a qualitative exploration of the plankton of a water. On hydrobiological excursions small casting nets, e.g. the Cosmos nets already mentioned give good service. With reservations, the plankton net is to be used if one wishes to carry out a continuous vertical catch from the bottom of a lake to its surface. A really continuous catch can in general be obtained only with a net. It is let down to the bottom and then pulled up slowly and evenly at a speed of about 10 cm/sec till it reaches the surface. By such a catch one can with reservations calculate the amount of plankton under 1 m^2 of lake surface.

The closing net

Often it is important to determine the exact distribution of plankton between the bottom of the water and its surface. This succeeds only if one makes catches periodically between defined depths in the vertical profile. The closing net has been constructed for this purpose; it can be closed under water by means of a drop-weight (messenger) (Fig. 26). If, for example, only the plankton between 20 and 10 m depth is to be taken, the closing net is let down to 20 m and then drawn up again slowly; at about 11 m the drop-weight is sent down and this

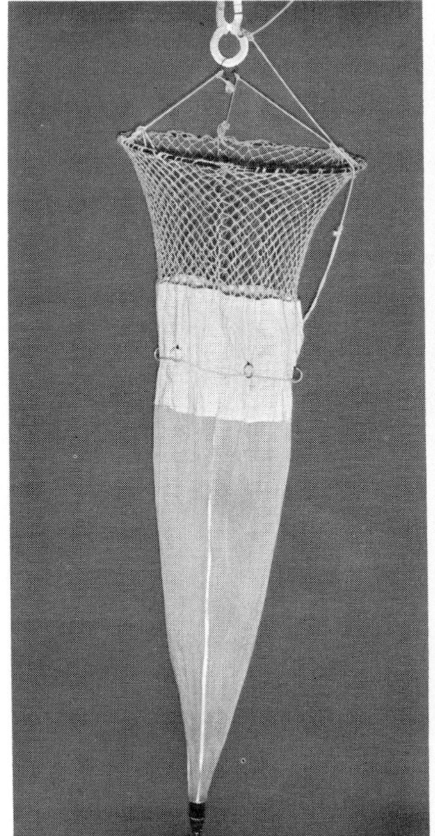

FIG. 26. The closing net, open.

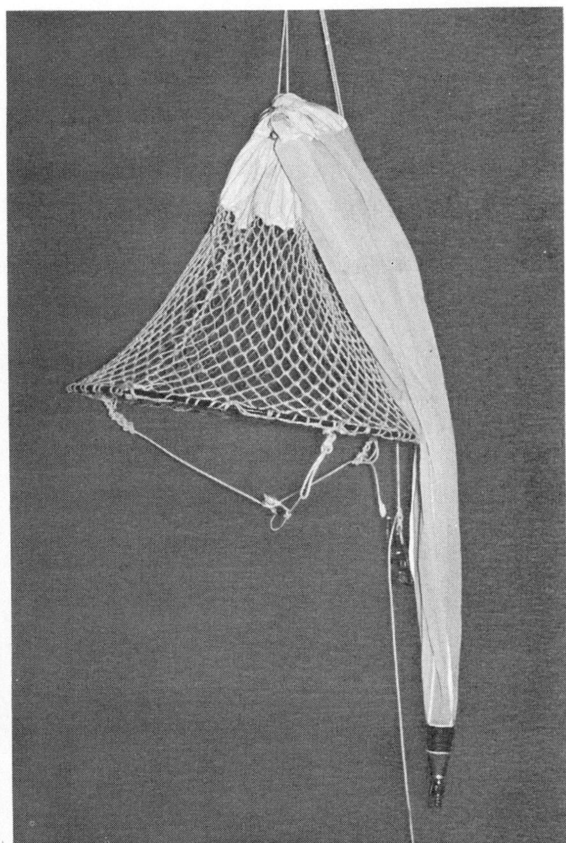

FIG. 27. The closing net, closed.

closes the net at 10 m. The drawing up of the net must not be interrupted for an instant. When the net is finally drawn up to the surface, no more plankton enters the closed net (Fig. 27); the animals collected come exclusively from the depth interval studied. In this way a lack of vertical distribution of the plankton can be discovered, but the variations which can be detected by the closing net depend on the order of magnitude of the net itself. Vertical variations of less than 1 m can no longer be determined with certainty. In general the intervals can be greater in the greater depths than they are nearer the surface. In the 252 m deep Lake Constance one would make catches at about the following intervals: bottom, 100, 100–50, 50–30, 30–20, 20–15, 15–10, 10–5, 5–3, 3–2, 2–1, and 1 m to the surface. In the choice of the intervals the denser population of the epilimnion, the vertical migration of the zooplankton, and the position of the thermocline and so forth should be borne in mind.

The closing net is made in the same way as the normal plankton net from which it differs only in having a closing mechanism. The bag of the net consists in its lower part of plankton gauze above which is a strip of coarse linen and a coarse network sewn in above the net ring (see Fig. 26). To the net rings are attached, as in the plankton net, three suspension cords tied to a small brass ring. This brass ring is attached above the closing mechanism to two cords—the draw-line and the closing line. The net is let down and pulled up by the draw-line; it should be marked in metres, and down it the drop-weight (messenger) falls. The closing line is led quite loosely round the part of the net made of coarse linen by means of rings sewn in. One end of it is firmly tied to one of the rings and the other is attached to the release mechanism. The net is closed in the following way. The drop-weight (messenger) sent down the draw-line strikes the release mechanism and frees the net from the draw-line. It now rests solely on the closing line which fastens like a ring round the bag of the net when it is drawn up, and this closes the net. At the same time, the upper part of the net tilts sideways so that no more plankton can get into the net (see Fig. 27).

As a release mechanism (Fig. 28), the closing apparatus of Auerbach has been especially well proven. It is made of brass. The individual parts of it are riveted in such a way that

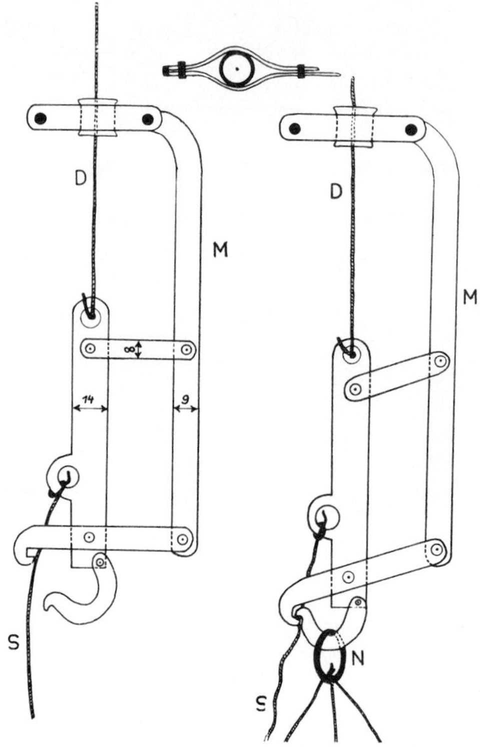

FIG. 28. Structure of Auerbach's closing device: on the right with an opened net; on the left with the net closed; the net hangs on the closing line S. D, the draw line; N, the ring of the closing net; M, the brass parts out of which the apparatus is made. The connections marked with a ● are fixed; those marked with a ⊙ can rotate.

they can turn on one another as far up as the guide ring for the line which is at the same time the abutment for the drop-weight. The closing net is hung on the hooks at N (Fig. 28), the drop-weight strikes the release mechanism and pulls down the brass bar M. This raises the lower cross bar, and the hooks bearing the net are disengaged. The net now hangs on the closing line and is drawn up by it as described above.

Other nets, used especially for marine researches, work with Schweder's release mechanism which closes the opening of the net itself. It consists of one or two (semi-) circular discs arranged vertically which come to lie, when they are struck by the drop-weight, horizontally, and close the net opening.

The Clarke–Bumpus plankton-sampler:

In order to meet the difficulty in quantitative evaluation of catches made with the net, Clarke and Bumpus (1950) constructed the horizontal closing net in the opening of which a screw propeller registers the amount of water passing through the net. By means of it the relationship between the amount of plankton and the volume of water fished through can be determined exactly.

The net lies horizontally in the water and works on the following principle (Fig. 29). The nets used are attached to a brass tube 12·7 cm in diameter and 15 cm long by a bayonet lock. The tube is so mounted in a frame that it is movable up and down so that in any position of the frame it lies horizontally. A metal bar fixed in front to a brass ring and behind to the net jar acts as a stabilizer to the net. The frame itself is so attached to the draw-line that it can swing freely and that the opening of the tube is directed forwards by any movement of it. Two vanes on the brass ring help to keep the horizontal position. In front the tube can be closed and opened by a cover which is pivoted like the damper in a stove pipe. The net is let down closed to the desired depth, opened with a drop-weight acting on the release mechanism, and, after the catch, is closed again by means of a second drop-weight. During the catch the propeller revolves in the tube and its revolutions are counted. Their number per unit time depends on the speed with which the water flows through. Because the amount of water that flows through also depends on the speed of the inflow, with a given net opening the number of revolutions of the screw can be calculated in water volumes if one has previously calibrated the apparatus. This calibration is made with a net fixed in a current canal and is done by the firm supplying the net. In general one revolution corresponds to 3·8–4·5 l. of water if one maintains a withdrawal speed of 1–7·5 km/hr. The error in the statement of the water volume is ±5%. At a lower speed, and when the net is more choked up, the propeller no longer works accurately. For the same reason only the net gauze with a degree of fineness up to No. 10 (see Table 3, p. 38) should be used.

Manufacturer: Mr. Fred Schueler, 80 Albemarle Road, Waltham 54, Massachusetts, U.S.A.

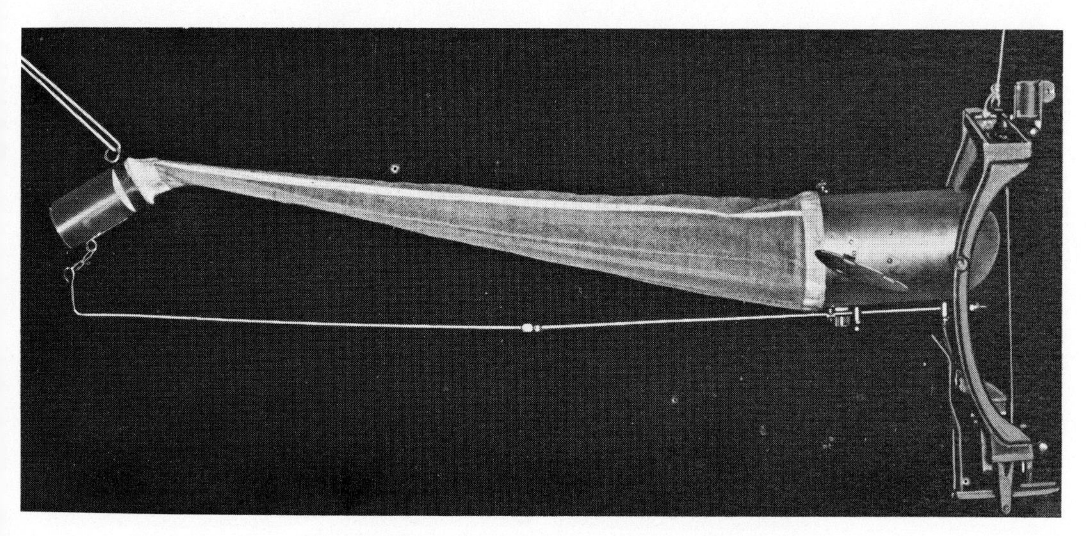

FIG. 29. The plankton-sampler of Clarke and Bumpus.

2. Collection of Plankton Samples with the Aid of Water-bottles

The water-bottle serves nowadays, especially in plankton research, for taking quantitative samples of plankton. It takes in an accurately indicated volume of water together with the plankton in it and has the advantage that by means of it the relationship between

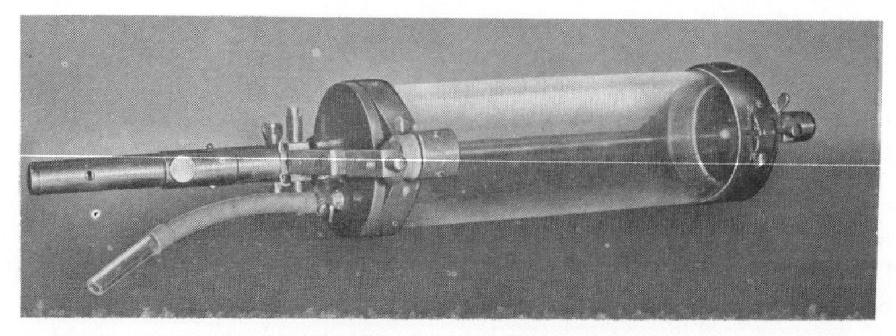

FIG. 30a. Friedinger's water-bottle closed.

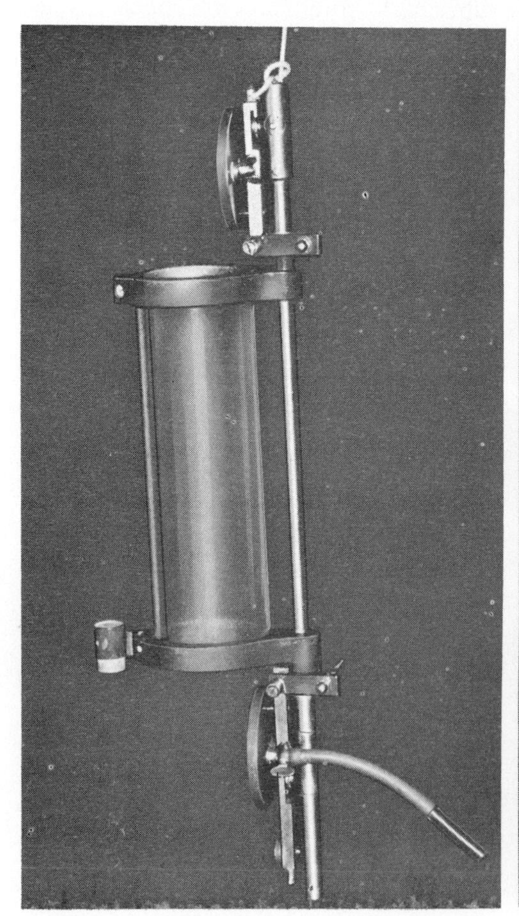

FIG. 30b. Friedinger's water-bottle, open with the lids upright.

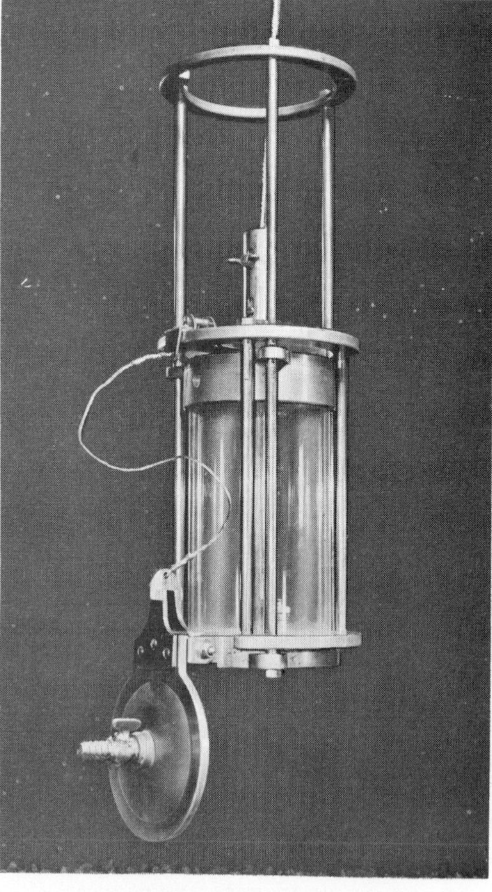

FIG. 31. Herbst's drop-bottle, open. (From Herbst, 1956.)

the volume of water and its plankton content is given directly. Elster (1958) has given a critical account of the use of buckets, and he established that buckets of the Ruttner type with horizontal covers to close them do not operate free from turbulence: when they are let down and water flows in, pressure waves and turbulence arise which disturb the plankton layers.

Bottles of the Friedinger type (Fig. 30) with vertical covers are better. These, like the Ruttner bottle, are closed and made watertight by a drop-weight with two hinged covers. Herbst has made a good model (Fig. 31) although it has a horizontal cover. The cylinder made of plexiglass (perspex or lucite equivalent) is moved by a drop-weight in such a way that it falls down for a short distance inside a sliding rail, and thus the column of water is, as it were, cut out and is almost simultaneously shut off (drop-bottle). According to the inventor the instrument operates largely free from turbulence and does not allow the larger plankton organisms time to escape. In comparison with the Ruttner bottle, therefore, 50% more copepods, for example, can be caught.

Horizontal water-bottles have been made by Wohlenberg and Schweder among others and by Jaag, Ambühl, and Zimmermann (see p. 14).

The water taken with a bottle, whatever its construction, is run out through a funnel on the outlet of which a piece of gauze is placed (only for zooplankton, not for phytoplankton). The piece of gauze is next transferred to a specimen bottle and fixed. For a quantitative estimation it is washed clean over a Petri dish.

Deficiencies of the water-bottle method

The bottles mostly have too small a volume and therefore give unreliable results. They must therefore be sent down several times to the same depths. The drop-bottle has a capacity of 2 l. there is also a 5 l. model of the Ruttner bottle, which, in spite of its less favourable construction, gave in the Lago di Varese relatively favourable results (Elster, 1958). In Lake Constance, Obersee, on the other hand, according to Elster, at least 30 and in the Untersee at least 15 l. at each level were necessary to obtain reliable results about the distribution of the Crustacea. Whether and how a bottle can be used must be decided by the density of the plankton estimated roughly by a catch with a net. When the density of the zooplankton is low a water-pump, which delivers at will large amounts of water from any depth, is to be preferred.

Compared with closing nets, bottles provide only random samples from a layer corresponding to the vertical position of the bottle. In this there is a risk that closely circumscribed plankton maxima are not taken or that their extent is overestimated, so that the values are too low or too high in the summation of the individual samples.

3. Collection of Plankton Samples with Water-pumps

The pump method has the advantage that as much water as one likes can be taken with it at a certain depth, so that the errors due to taking too little water can be excluded. In order to obtain a constant suction action it is best to use the electrically (accumulator) operated rotary pump made by Messrs. Allweiler which has a flow meter with which the amount of water pumped can be directly measured. A soft rubber or plastic hose is attached to the pump, the open end of which is let down to the depth under investigation. The water that passes through it into the hose must then be pumped out. Then, according to the density of the plankton, 30–50 l. of water are pumped up and filtered through various fine gauzes according to what one wishes to retain. If one uses finer nets than No. 16, the phytoplankton counting interferes with the subsequent counting. The plankton is fixed and kept for further treatment.

Catches should not be made with the pump in order to obtain living material for cultures or physiological investigations because uncontrollable damage may be done to the organisms by the pump.

An important advantage of the pump in comparison with nets or bottles is the fact that with the pump one can take a vertical profile at intervals as small as one pleases. This is especially important in studies of the uppermost layers of the epilimnion and in the thermocline where variations in the vertical distribution of the plankton occur at especially narrow intervals.

Sources of error in the pump method

Researches have shown (Herbst, 1957; Elster, 1958) that the larger plankton organisms, especially copepods, react by flight when they get into the suction area, and for pump collections these reactions must be borne in mind. One can lessen the risk thus created of taking selective catches, if one connects the tube to a funnel which hinders the escape of the plankton organisms. This precaution is, however, necessary only for tubes with diameters up to 2·5 cm. Tests have shown that with tubes with diameters between 2·5–4 cm, and with the use of the hand-operated Allweiler rotary pump No. 2, the percentage of plankton organisms escaping has practically no significance (Elster, 1958). Under these conditions catches made with the pump give representative quantitative results, so that they can be used for the calibration of nets and bottles. A committee on methods of the International Association for Theoretical and Applied Limnology (IVL) is also concerned with these problems.

III. FIXATION AND PRESERVATION OF ZOOPLANKTON

For quantitative estimates the zooplankton is fixed and preserved with a formalin (= formaldehyde) solution of about 2–3%. Commercial 40% formalin is suitable for this and a correspondingly small quantity of this is added to the sample. Because many plankton organisms, especially Cladocera, remain clinging by their shells to the surface film, it is very advantageous for subsequent counting to lower the surface tension of the water by means of a small dose of liquid detergent. It is simplest to add a little of the detergent to the formalin used for preservation. Many species lose, when they are preserved, their characteristic shape and are more or less contracted. This, however, hardly interferes with the quantitative enumerations if one knows the change of shape of each species; one can recognize them all the same.

For determination of species, on the other hand, more accurate methods of fixation and preservation must be considered for the different groups of animals and plants. These are summarized in Appendix I, pp. 159 ff. A simple method of separating the zoo-net plankton into the most important groups has been described recently by Straškraba (1964).

IV. ENUMERATION OF THE ZOOPLANKTON

The objective of quantitative study of plankton is to determine the density and the composition of the plankton referred to a unit volume of water. The quantitative catches made for this purpose must naturally also be strictly quantitative. To ensure this it is necessary to find out accurately the number of organisms in the individual catches. Because it is usually a question of very large numbers of organisms, all of which can never be counted in a reasonable time, one contents oneself with a part of the sample. The art of it lies in so separating off this part that it has the same composition and is representative of the whole sample. Only under these conditions can the number of organisms in the whole sample be calculated from the results obtained with a part of the sample. Thus the count of a plankton sample proceeds by the following stages:

1. Making a suitable concentration of the plankton and taking a sample part of it.
2. Placing the plankton in a counting chamber.
3. Counting this.
4. Calculation of the plankton content of the whole sample.

1. Preparation of a Suitable Plankton Concentration and Taking Samples for Counting

The quantitative plankton sample taken with a net, bottle, or pump and preserved in the Pril-formalin solution is next quantitatively poured into a measuring cylinder. If the sample is poor in plankton, water is poured off after the organisms have sunk to the bottom. Samples very rich in plankton may, on the other hand, be diluted. The final volume of the sample is accurately noted and it is put into a shaking vessel of corresponding size. The measuring cylinder is rinsed out and the water used for this is added to the plankton sample and is also put into the shaking vessel. The volume of fluid in the shaking vessel must be accurately known. Simple round flasks with a wide neck closed with a perforated cork serve as shaking vessels. Into the hole in the cork a piston-pipette (or Stempel pipette) is introduced and let down until its lower end at least reaches the middle of the plankton sample.

Hensen's model of this (Fig. 32) is best suited for this purpose. When the piston is drawn up the plankton sample is caught between the sliding piston and the pipette tube. This happens whilst the sample is shaken and equally distributed in the shaking vessel. Piston-pipettes are available for various volumes between 0·1–5 ccm, but they are, because of their accurate manufacture, pretty costly.

In the U.S.A. instead of the syringe the "plankton subsampler" is used and this also makes it possible to take a representative counting sample. This is done by distributing the whole sample equally over a surface which is divided into areas of equal size. The content of one or more of these areas is then further treated as a counting sample. These "surface area" methods have the advantage that the subsampler can easily be made up if one subdivides appropriately a glass trough.

Literature: Cushing, 1961; Hopkins, 1962.

2. Filling the Counting Chamber with Plankton

In normal counts of crustacean plankton taken from lakes about 1 ccm is taken up with a Stempel pipette and quantitatively put into a counting chamber. Rinsing out is not necessary because the piston is not wettable, but one must take care that no plankton remain, adhering to the outside of it.

The counting plate is either a simple glass plate with a surface of about 8 by 4 cm or, better, a glass plate to the edge of which strips of glass are cemented to make a counting

FIG. 32. Piston-pipette (Stempel pipette) and shaking vessel for zooplankton. (Photo by Hydrobios.)

chamber. A sloping edge of the glass strips makes possible good microscopical observation up to the edge. In both instances the bottom of the plate is divided into longitudinal strips which can easily be marked on the glass with a diamond. It is best to mark out on plexiglass, (perspex, lucite) which is easily marked, strips 0·5, 1·0, and 1·5 mm broad.

The counting sample is put on the counting plate with the pipette, care being taken that the layer of it is not too thick. The concentration of the plankton sample in the shaking vessel made at the beginning should be suitable in relation to this. In a layer that is too thick, several plankton organisms lie over each other and too close to one another and can then be recognized only with difficulty; also too much phytoplankton causes difficulty. Another source of difficulty is the shaking when the plate is moved to put it under the microscope. To avoid this shaking completely, one puts the counting sample into a counting chamber instead of on to a plate, this chamber being filled up to its border and closed with a cover-glass. To fill up the chamber one takes either more of the counting sample or, if this is too rich in plankton, water free from plankton.

3. Counting the Plankton

The counting plate or chamber is counted under a microscope, preferably a stereomicroscope, at a magnification which permits the identification of the organisms with certainty. A mechanical stage is to be recommended because it permits even movement of the chambers without shaking. Species or groups of species are counted according to the problem involved, or age groups of the individual species are differentiated. Thus one counts, for example, rotifers, Cladocera, calanoid or cyclopid copepods, etc., or one subdivides these groups and counts among the Cladocera the *Bosmina* and *Daphnia* (Fig. 33) and sometimes even distinguishes between the various species of both these genera. Very often (in researches on population dynamics) it is important to count separately, among the copepods for example, the adult males and females as well as the developmental stages. How far one will and can subdivide in this way depends on the research planned.

It is self-evident that before beginning the actual count one takes a qualitative sample to find out what is to be expected. Then before the count one makes out lists of the various categories in the count. For each single animal counted one makes a vertical stroke in the corresponding category, and for each fifth animal a transverse stroke through the four group, so that the groups of five can later be quickly added up. One can also use mechanical counters which by pressure on a knob continue counting and immediately give the final sum; naturally a counter of that kind must be ready for each group counted.

When the whole chamber has been counted it is emptied, carefully washed out, and, if necessary, filled again. The result of the count of each chamber is noted separately; only in this way can the counts made on successive fillings of the counting chamber from the same sample be compared with one another. The protocols of the count must naturally be

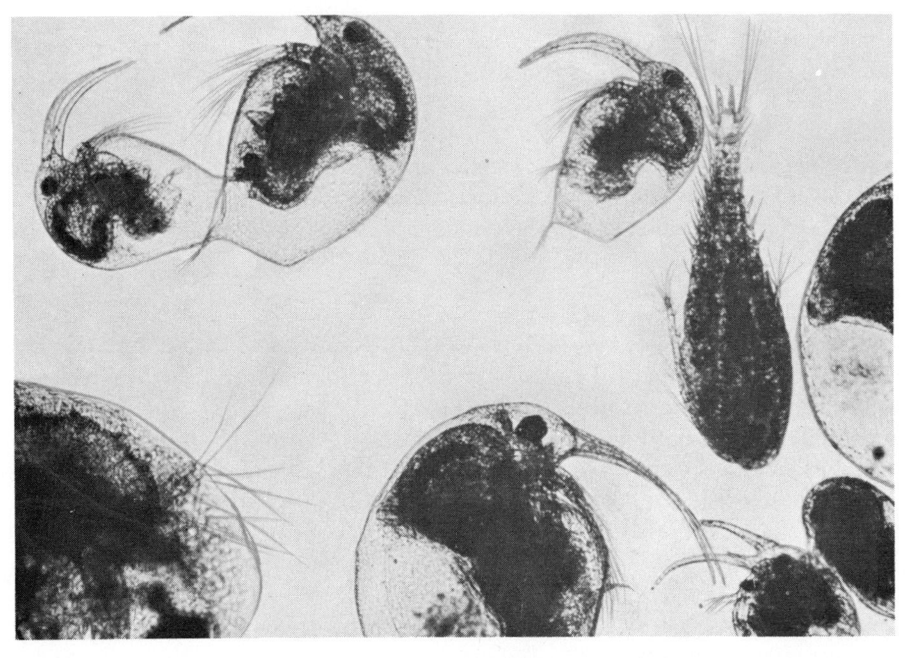

FIG. 33. Sample of zooplankton, chiefly consisting of bosminas: on the left below, part of a Daphnia; on the right, a copepod. (Magnified about 50 times.)

kept during the whole investigation because they constitute a scientific document and possibly must be consulted for comparison in later work.

If the plankton sample counted is to be kept further, it is preserved again in formalin solution or, better, in alcohol to which glycerine has been added (cf. methods of fixation and preservation in Appendix II). Alcoholic samples may, however, be difficult to count.

4. Estimation of the Plankton Content of the Whole Sample

Because the count records are the basis for the later calculation of the total plankton content of the sample and of its composition in the lake or in the different parts of the lake, a foundation must be made by careful and extensive counting on which mathematical calculation can be based with a good conscience. Errors made in the foundations will add up in every addition, multiplication, etc.

The question of what percentage of the whole sample is to be counted depends on its plankton content or on its concentration in the shaking vessel and its species composition. Because the volume of the sample cannot be as small as one pleases, the concentration of samples poor in plankton cannot always be conveniently standardized; if necessary the whole sample must be counted. If the results of the counts made on the various fillings of the chambers no longer differ essentially, the added values of the count can be checked statistically and considered as being representative of the whole sample. If there are rare species, one can count these in additional chambers and disregard all the others.

If one has counted about 5% of the total volume of a plankton sample and has convinced oneself that this portion represents a statistical average, then the plankton content of the whole sample is obtained by conversion to 100% (i.e. multiply here by 20).

This final value is the amount of plankton in a catch made with the net, bottle, or pump and is accordingly related to a certain volume of water. If, for example, it is a question of a vertical catch from the bottom to the surface with the normal net ($\frac{1}{4}$ m^2 net opening), the plankton value obtained by counting and conversion is 100%, and, multiplied by 4, gives the plankton content of water column under 1 m^2 of lake surface. A 5 l. bottle catch must be multiplied by 200 and a 20 l. pump catch by 50, if one wishes to calculate the plankton content in each cubic metre and so on. If one has made catches at intervals with the closing net or other apparatus, the plankton content of each unit of volume at the individual depth intervals can be compared with one another. Once again it may be emphasized that the errors made in both the catch and in the evaluation of the plankton sample will be multiplied by the same factor.

PHYTOPLANKTON

The phytoplankton consists almost entirely of algae, most of which are very much smaller than the plankton animals. Figure 34 gives a good comparison of their relative sizes although it shows large plankton algae. Only plankton algae which form cell colonies, such as *Volvox* and some others (Figs. 35 and 36) approach the size of the zooplankton.

This causes special difficulties in the study of the phytoplankton which have been overcome by the development of the inverted microscope, by special counting chambers and counting procedures and also by the abandonment of the use of nets. These procedures are given here together, although they show many parallels with the methods for zooplankton. The phytoplankton is divided into three size classes (cf. also p. 33):

1. The net plankton, the largest forms which also can be caught with fine-meshed nets (gauze No. 25, diameter of the pore 60 μ), such as *Asterionella*, *Tabellaria*, *Fragilaria*, *Volvox*, *Oscillatoria*, *Ceratium*, *Peridinium*, etc. (Figs. 2 and 37).

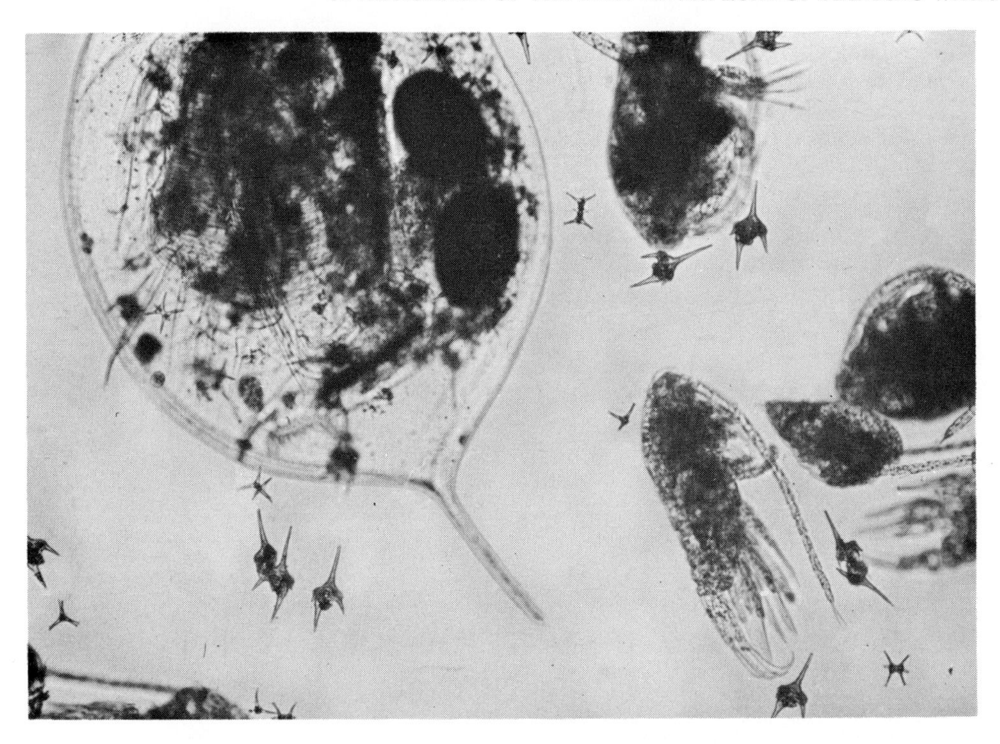

FIG. 34. Comparison of the sizes of zooplankton (daphnias and copepods) and phytoplankton (*Staurastrum*, *Peridinium*, and *Ceratium*). (Magnified about 60 times.)

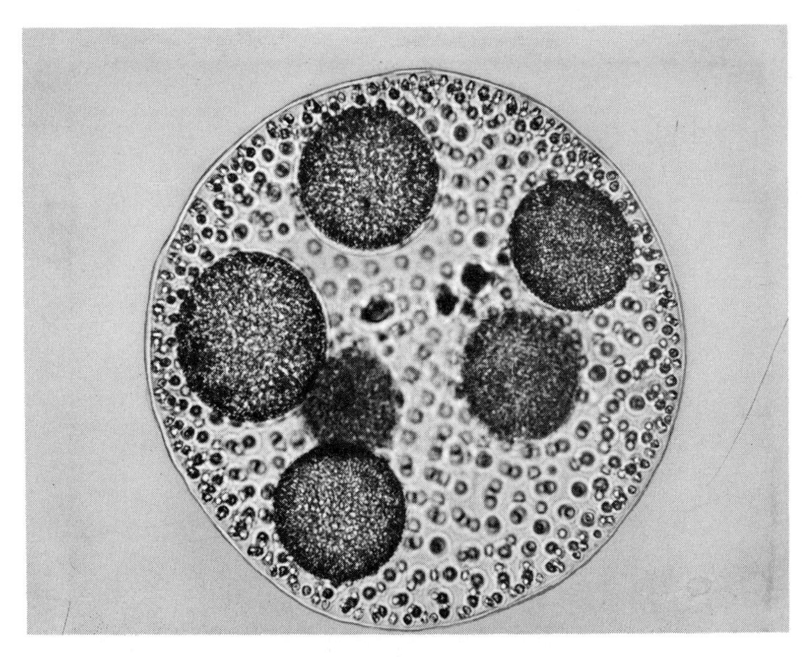

FIG. 35. Cell colony of a *Volvox* alga, one of the largest phytoplankters of fresh water. (Magnified about 100 times.)

2. The nannoplankton, small algae less than 60 μ, which cannot be caught with the finest nets.
3. The μ-algae (ultraplankton), the smallest forms of all, measuring only a few microns.

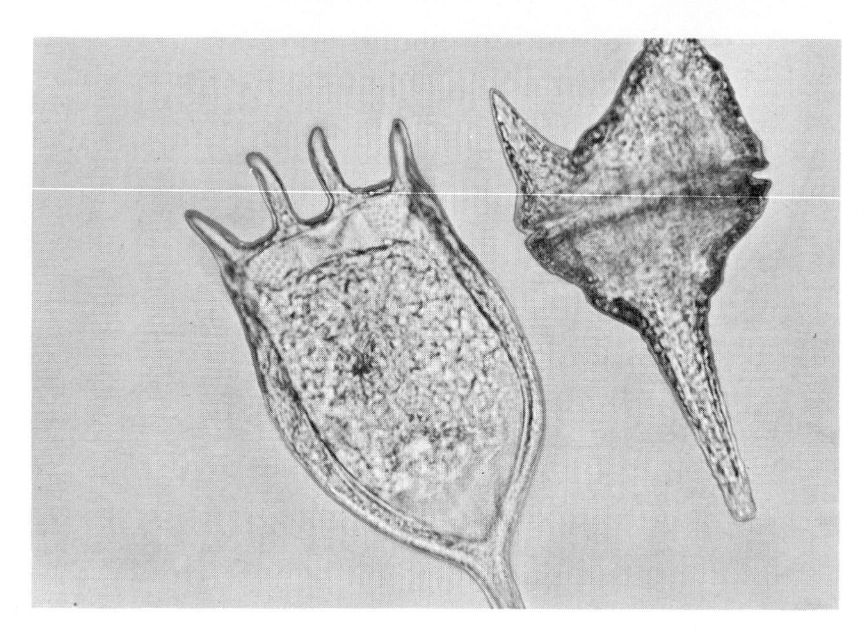

FIG. 36. The largest phytoplankter is about the size of the smallest zooplankter. On the right, the dinoflagellate, *Ceratium*; on the left, the rotifer, *Keratella*.

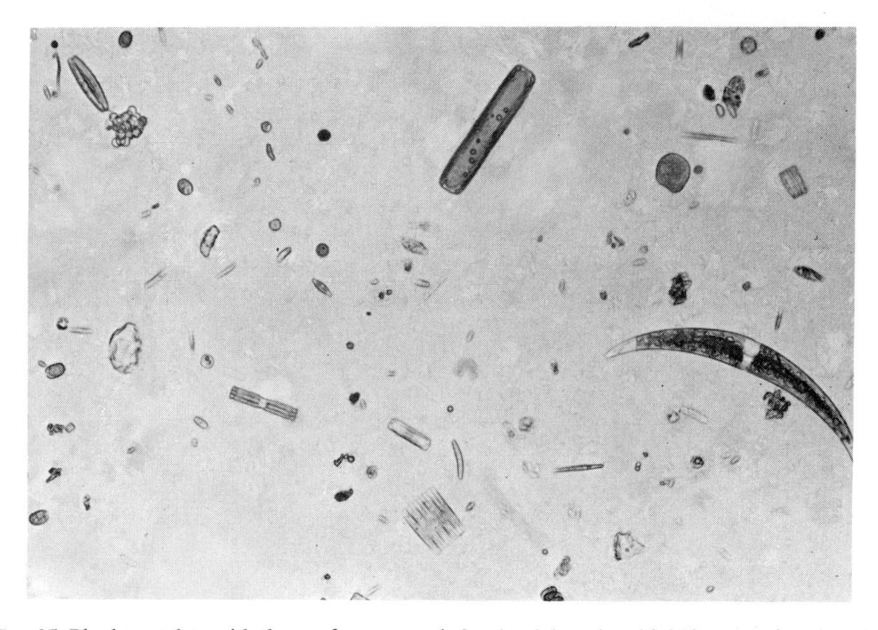

FIG. 37. Plankton taken with the net from a pond. On the right a desmid *(Closterium)*; otherwise chiefly diatoms.

From these size classes it follows that neither the nannoplankton nor the microplankton would be caught by nets. However, the recent researches of Rodhe (1958), Nauwerck (1963), and others have shown that the nannoplankton plays an important part in primary production and in the nutrition of zooplankton in the lake. Any use of nets is, for this reason, excluded from quantitative phytoplankton methods. Nannoplankton and μ-algae can only be concentrated by sedimentation and centrifugation of water taken with bottles or pumps.

I. QUALITATIVE INVESTIGATION OF THE (NET) PHYTOPLANKTON

Knowledge of the net plant plankton is important for establishing the character of a piece of water. Oligotrophic and eutrophic lakes show very different plankton pictures. Also the composition of the plankton changes in the course of the year in the same lake, as Figs. 38 and 39 of the Great Plöner Lake in Holstein show.

The phytoplankton net is therefore still an important limnological instrument. Because it must be provided with gauze No. 25, its open sieve surface (cf. p. 37) is only about 20% of the total surface and its filtration resistance is correspondingly high. This causes an intense progressive blocking of the net. A smaller net opening therefore greatly extends the filtration surface, and in this way a very long and narrow net is formed, the zeppelin net (Fig. 40).

The zeppelin net is conical at its lower end, the upper part having a cylindrical extension of net gauze so that a large net surface is achieved. The net used in our institute is 1 m long and has a net opening with a diameter of 15 cm. To the uppermost ring of the net a strip of strong linen 6 cm broad is sewn. Two other metal rings are sewn into the net itself, one between the cylindrical extension and the conical part and the other in the middle of the cylindrical extension (Fig. 40). This ensures that when the net is pulled slowly through the water it always remains open.

It is easy to calculate that the net described has a gauze surface of 4000 cm^2; the net opening measures 180 cm^2 which corresponds to a relationship of 1:22 (in a normal net made of gauze 8 for zooplankton it is 1:4). Because, however, the effective mesh surface of this gauze amounts to only about one-fifth of the gauze surface, the actual relationship of the net opening to the real filtration surface is reduced to 1:4·4. The zeppelin net therefore has no noteworthy filtration resistance in pure water. However, in water rich in plankton a very rapid blocking of the net must be expected, so that the filtration resistance quickly increases and the large species can no longer be quantitatively retained. The zeppelin net is therefore suitable only for qualitative exploratory catches and for capturing net plankton for cultures, indentification, demonstration, etc. Recently gauzes—the pores of which have a diameter of about 30 μ—have been manufactured for the capture of nannoplankton (the nannoplankton net of Höll, made by Messrs. Bergmann), but this retains only the largest nannoplankton and it is not suitable for quantitative work.

II. QUANTITATIVE INVESTIGATION OF THE PHYTOPLANKTON

Nets are never used for this, only water-bottles, by which smaller amounts of water suffice than when zooplankton is taken. The Ruttner and Friedinger types of water-bottle, as well as other types, can be used. The method of using these is described in the hydrographic and zooplankton sections. Catches made with the pump are as good as those made with the bottle but are not necessary because of the small amounts of water needed.

FIG. 38. Winter plankton from the Great Plöner Lake, chiefly filamentous Oscillariaceae: below two *Ceratium*; and above them a *Keratella*. On the left beside them a *Staurastrum*.

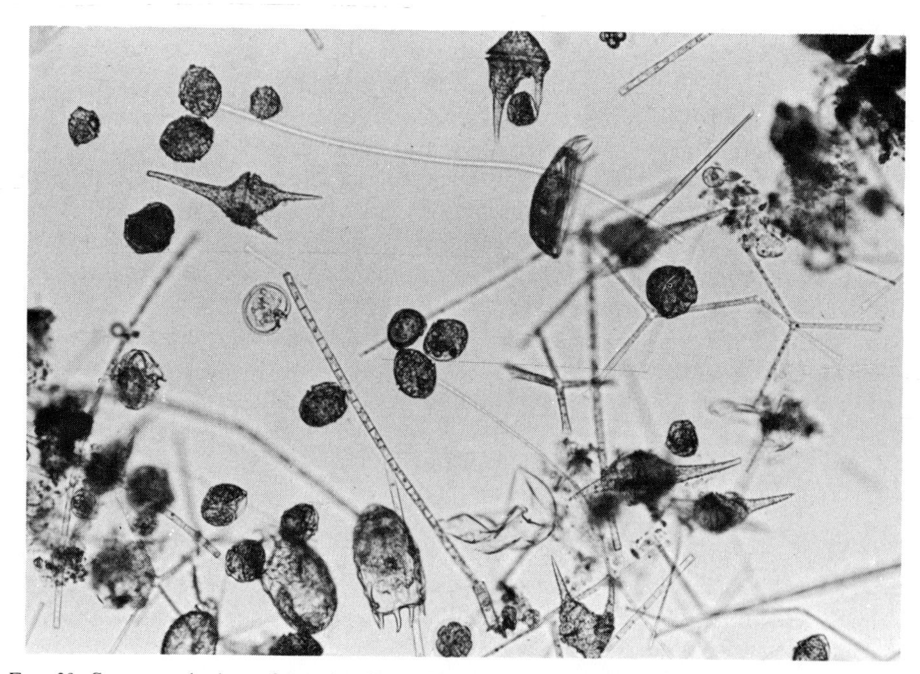

FIG. 39. Summer plankton from the Great Plöner Lake: *Peridinium*, *Ceratium*, *Keratella*, and *Diatoma*; practically no Oscillariaceae.

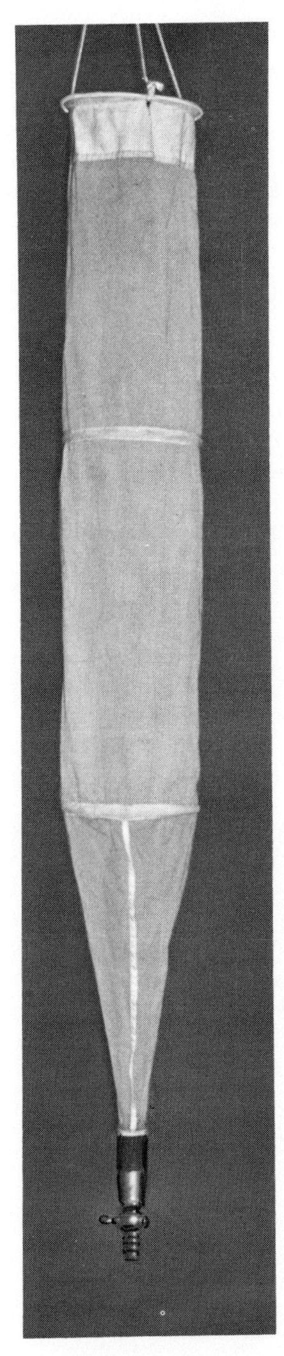

FIG. 40. The zeppelin net.

1. Fixation and Preservation

About 100 ml of the contents of the water-bottle are put into an appropriate sample-bottle provided with a ground-glass stopper (or a screw-top covered with polyethylene) and 2–3 drops of concentrated aqueous solution of iodine in potassium iodide (Lugol's solution) plus 10% acetic acid or sodium acetate are added until the fluid has the colour of cognac. The iodine fixes and preserves the organisms and at the same time colours them; the acetic acid preserves the flagella and cilia. Treatment with Lugol's solution has, however, a further decisively important advantage: by taking in iodine the organisms become heavier and sink more quickly to the bottom. This sedimentation of these is absolutely necessary before they are counted. The sample thus treated is kept tightly closed up in the dark and then will keep for years.

To make Lugol's solution

Dissolve 10 g of neutral potassium iodide, analytical quality, in 20 ml of water and add to it 5 g of twice sublimed iodine of analytical quality. When solution is complete add 50 ml of water and 5 g of sodium acetate of analytical quality or, instead of the sodium acetate, 10% acetic acid. Store in a 100 ml narrow-necked flask made of neutral glass (e.g. Jena laboratory glass) with a ground-glass stopper that fits well.

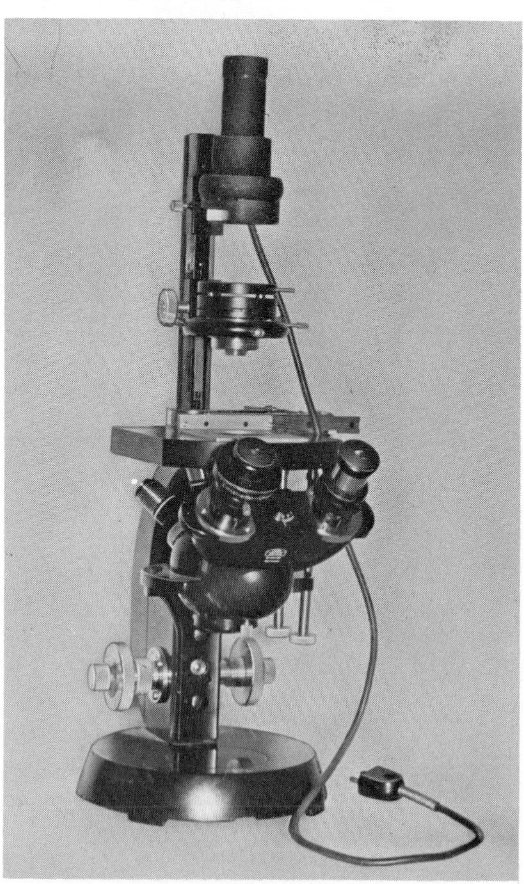

Fig. 41. Utermöhl's inverted microscope.

2. Counting the Phytoplankton

Principle. *For various reasons phytoplankton can be counted only with difficulty under a normal microscope. The plankton sample must be counted in a so-called tube chamber, on to the cover-glass of which the plankton sinks in a certain time. The supernatant water, however, cannot be decanted and remains in the chamber. When a normal microscope is used the necessary high magnification would immerse the objective in the water, and the use of a water-immersion system gives optically poor images.*

For these reasons Utermöhl's (1928) inverted microscope is used in which the objective is directed from below up to the bottom plate of the counting chamber so that the plankton lying on the cover-glass can be examined without difficulty, even with immersion lenses (Fig. 41).

Counts made on plankton samples fixed with Lugol's and acetate proceed by the following stages:

1. Filling the plankton chambers and sedimentation.
2. Putting the sedimented samples on the microscope.
3. Actual counting of the samples.

Filling the plankton chambers

Kolkwitz (1907, 1911) has developed the procedures with chambers. He used, however, only very flat chambers containing 0·5 ml (Fig. 42) with a shallow sedimentation depth which were counted with the normal microscope and viewed from above and leaving out

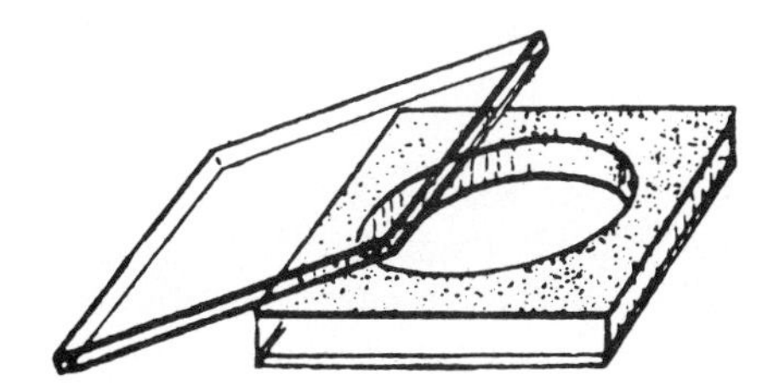

Fig. 42. Kolkwitz's shallow chamber. (From Ruttner.)

the identification of the smallest forms. It is a disadvantage that there is only a thin layer of plankton in these flat chambers and that the surface examined, on the other hand, is relatively large. The Kolkwitz chambers can, however, still be used if an inverted microscope is not available, especially if the plankton sample contains a great deal of plankton. Kolkwitz chambers that are also suitable for the Utermöhl microscope are also commercially available (Fig. 43). They are hollowed-out plexiglass (perspex) plates containing 0·5 or 1 ml.

Kolkwitz chambers are also very suitable for the study of plankton from small collections of water (pools and ponds). The chambers can be closed directly underwater with their coverglass. Rylov (1927) has described a simple apparatus with which Kolkwitz chambers can be filled and closed as far down as depths of 1 m.

Modern phytoplankton methods use the so-called tube chambers (Utermöhl, 1932). These are cylinders which, in contrast to the Kolkwitz chambers, are taller than their

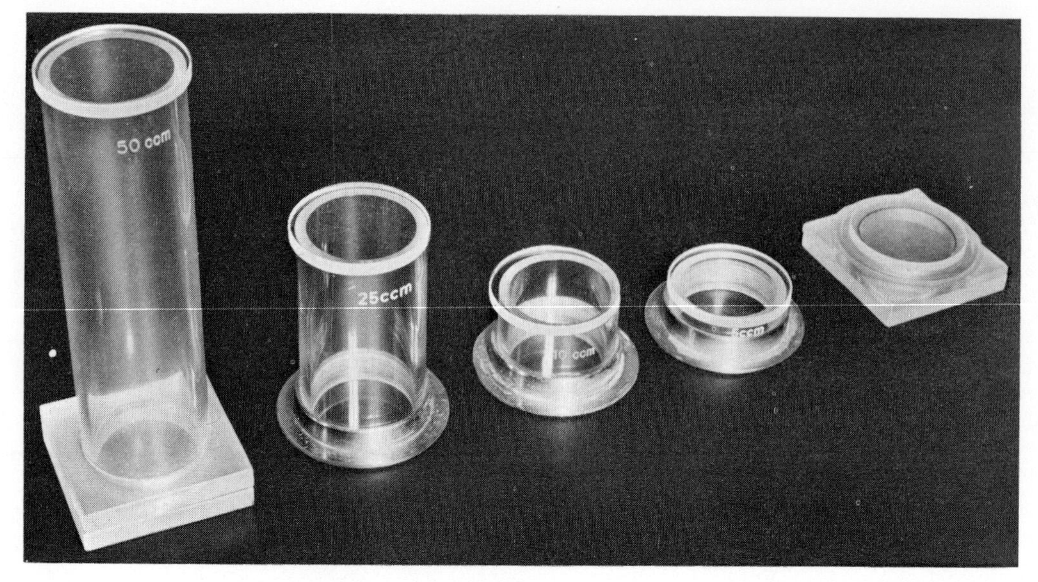

FIG. 43. Different tubular chambers of Utermöhl. On the left a compound chamber (see text).

diameter (Fig. 43). With them a longer sedimentation is possible, but also a higher concentration of the plankton reaches the bottom plate, so that the rarer species are also detected. These chambers are especially advantageous or even necessary when the samples are poor in plankton. Chambers of differing capacity are selected according to the density of the plankton, but the size of the bottom surface remains constant.

Making tube-chambers

These are nowadays made, as the Kolkwitz chambers are, out of plexiglass (perspex, lucite) (Lund, 1951). It is easy to make them out of tubular pieces of plexiglass or of some equivalent synthetic material by cutting this to a suitable length and grinding the ends smooth. Their size is optional; in practice sizes suitable for the content are chosen. A circular microscopical cover-glass that fits the lower end is cemented over this end and on it the plankton is later collected. Utermöhl recommends as a suitable adhesive the Araldite made by Messrs. Ciba of Basel. The chamber is closed above by a loose circular or square glass-plate.

The counting of the plankton is a question of time. The result must bear a reasonable relation to the time required to achieve it. One does not count the whole sample any more than one does with zooplankton, but stops counting when there is a representative result. Many authors have been concerned in detail with the question of when a result of the count can be regarded as significant (Lund, Kipling, and Le Cren, 1958; Javornicky, 1958; Nauwerck, 1963).

The plankton content of individual samples can vary a great deal, so that chambers of different sizes must be used. The richer in plankton a sample is, the smaller the chamber may be. Careful choice of a chamber makes it possible to obtain an even layer of plankton on the bottom of the chamber that is suitable for counting. The following is a good rule of thumb: if there are more than 10,000 plankton organisms in 1 ml one uses flat chambers (Kolkwitz; flat tube-chambers); if they are less numerous than this, one uses tubular cham-

TABLE 4

Volume (ml)	Internal diameter (mm)	Content (mm^2)	Height (mm)
1	10	80	25
5	16	200	25
10	18	255	40
20	20	315	64
50	25	500	100
100	34	910	110

bers or compound chambers. It is best to make or buy one or more sets of chambers of different sizes. Utermöhl (1958) recommends the volumes and sizes shown in Table 4.

It is, however, good practice to keep the internal diameter of the chambers, and therefore their bottom surface, constant, and to change only their heights and therefore their volume. Then the same correction factors (see Table 2) apply to partial counts. Messrs. Carl Zeiss (Oberkochen) supply tubular chambers of different heights, but all of them have a diameter of 25 mm (Fig. 43).

To fill the chambers they are put in a Petri dish. The flask containing the fixed plankton is very well shaken to distribute the plankton evenly, and the tube chamber is filled to overflowing. It is important beforehand to bring the plankton sample to room temperature so that no air bubbles are formed in the tube chamber during the sedimentation. The cover-plate also must be put on without air bubbles. The chamber is then dried and set aside without shaking until all the plankton has sedimented down. The time needed for this varies with the height of the chamber.

Nauwerck (1963) found that the relation of the height to the diameter influences the sedimentation. If the height exceeds 5 times the diameter the convection currents are so strong that significant amounts of plankton do not sediment at all. In chambers of suitable size 4 hr are needed for each centimetre of height to sediment the smallest organisms.

The compound chamber of Utermöhl (1952) is a combination of the Kolkwitz and the tubular chambers after the suggestions of Kolkwitz (1932) and Lund (1951). On top of a shallow tubular chamber of 2 cm^3 capacity with a sedimentation height of 4 mm, inserted into a square plexiglass of the same size, there is a tubular chamber open below and also inserted into a plate of the same size and so placed that it is watertight (Fig. 43, left). Then the tubular chamber is filled as described above and it is set up so that it is free from shaking. To ensure a quantitative sedimentation the tube on top should not be too high. The advantage of these compound chambers is that the total volume of the plankton sinks down into the shallow plate chamber. It is now necessary to slide the top tubular portion from the plate chamber in such a way that none of the plankton is stirred up and so lost. To do this one places on the right and left of the plate chamber two plexiglass plates of exactly the same thickness and surface area and holds one of them steady with one hand and with the other hand slides the tubular portion together with the cover-glass of the plate chamber on to the plate held steady. The cover-glass slides back and accurately closes the plate chamber, which is thus ready for counting on the inverted microscope. This sliding operation is easily done with accurately made chambers.

Small irregularities on the edge of the plates easily causes them to interlock, and these can be removed by quite light filing.

The count itself

Counting phytoplankton is more difficult than counting zooplankton. It requires much higher magnifications and therefore more time. The chambers put on to the counting microscope are examined from below through the cover-glass bottom and counted. The choice of the magnification is decided by the smallest of the organisms; for the especially important nannoplankton a magnification of $\times 1000$ is necessary. For this reason the time needed to count a plankton chamber is increased to quite an extraordinary degree. It is not rare to find in eutrophic lakes 1 million plant cells per millilitre. One can thus get an idea of the duration of the count itself without regard to the pains which must be taken to identify the smallest forms.

Because one can never devote one or more weeks to the counting of a plankton chamber, one must restrict the count to a part of the bottom area of the chamber.

It is best to count two strips perpendicular to each other, over the middle of the bottom of the chamber. For this an ocular is necessary marked with two parallel threads the distance of which from each other can be adjusted. Transversely across them is another thread. The ocular is put into the microscope in such a way that the parallel threads lie in one of the two directions in which the mechanical stage moves. This is done by seeing that the plankton distributed through the field of vision does not cross the parallel threads. Only the strips between the parallel threads are counted: and because the device can be adjusted at will, wide strips are selected when the plankton is less dense and narrower when it is denser. The effect of the magnification must also be remembered. The higher this is, the narrower is the breadth of the visual field. Because all the strips counted must be within the field of vision, the breadth of them cannot be selected at will.

By means of the mechanical stage all the strips to be counted are passed through the field of vision, and everything between the threads is counted. A plankton organism is counted if it is displaced beyond the thread that is perpendicular to the threads marking out the strips being counted. Also included in the count are any plankton organisms which lie on one of the two threads or half inside and half outside the strip being counted. But in deciding about this one should consider only either the left-hand or right-hand thread. If one wishes to count other parts of the bottom of the chamber, one makes a displacement of one whole counting strip at a time, and thus the whole chamber can be counted under certain circumstances; or one counts two strips of the same size perpendicular to one another over the greatest breadth. In this way irregularities in the distribution of the plankton over the bottom of the chamber which arise during sedimentation are levelled out. One can also count several strips which lie at an angle of 45° to one another (Utermöhl's diametric count). It is best to use for this the chambers made by Messrs. Zeiss (Oberkochen) with a diameter of 25 mm. Because the four strips to be counted then add up to a length of 100 mm and the bottom of the chamber measures 500 mm^2, the conversion to the total plankton content of the chamber or to 1 mm^3 is very simple.

The decision about which part of the surface will be counted is made according to the frequency of the individual species in the sample. Utermöhl (1958) suggests a count of at least 100 individuals of each of the more important species in the sample in order to obtain a sufficient degree of statistical accuracy. Nauwerck (1963) had at times, in the Swedish Lake Erken, to deal with 30–40 species, of which he counted 1–3 species in the order of magnitude 10^2, 10–20 species in the order of magnitude 10^1, and the remainder in the order of magnitude of 10^0. The sizes of the counting chambers were chosen according to the concentration of the plankton in such a way that the frequency of the species corresponded to the order of magnitude mentioned. In this way the time needed for counting a sample could be reduced to $\frac{1}{2}$–2 hr. It is advisable to use several chambers for each sample: Utermöhl (1958) suggests the following kinds of chamber for eutrophic waters:

1. The 50 ml tubular or compound chamber for the rare species.
2. The 10 ml tubular or compound chamber for species occurring with average frequency.
3. The 1 ml tubular chamber for the species that are numerous.
4. The flat chamber for species that are very abundant.

Flat chambers can generally be left out for oligotrophic lakes. For the most accurate work it is necessary to use as controls at least 2–3 equivalent chambers for each frequency class, and for all the frequency classes 6–9 chambers for each sample. The degree of accuracy in the quantitative evaluation and the amount of apparatus used is conditioned, as it is in all investigations, by the problem in hand. Lund *et al.* (1958) and Javornicky (1958) arrived at the following mathematical formula for the estimation of the statistical significance of the results of the count:

$$f_{max} = \pm 2 \frac{100\%}{\sqrt{n}},$$

that is to say that with a numerical value of 25 cells, the range of error is $\pm 40\%$; when $n = 100$ this error falls to $\pm 20\%$ and when $n = 10{,}000$ to $\pm 2\%$. It is customary to calculate the results of the count to the number of cells per millilitre or litre of lake water. If one has obtained in a chamber with a diameter of 25 mm = 500 mm² bottom surface, a result of the count Z, then $Z \times 50$ is the total result for the chamber; Z can also be the result of the count for an individual species. The total result for the individual chamber, divided by its volume, gives the number of cells per millilitre for each species counted. Addition of all the individual values gives the total number of cells per millilitre of lake water or by multiplication by 1000, the total number per litre of lake water. These values can be directly compared with one another.

3. Other Methods of Evaluation

It is not only always a question of the exact number of cells and of the distribution of species in the lake, but often also it is a question of determining the biomass of the phytoplankton. To do this one must know the weight of the individual algae. With zooplankton it is relatively simple to determine the fresh and dry weight of several members of a species. With phytoplankton this is not so, because one can never capture pure larger quantities of a species; always a mixture of plankton, mixed with floating debris and zooplankton, is taken. For estimations of the dry weight that are sufficiently accurate, pure cultures of the various species would be required, and these have so far been successful in only a few instances (cf. Appendix I). For the same reasons the estimation of the sedimentation volume suggested by Apstein (1896) cannot be used.

The method of Lohmann (1908) gives approximately correct results. The volume for each individual species is determined by measuring the displacement volume of full-scale models, or the species are reduced to easily calculable, simple geometric bodies and their content calculated. One can take the specific gravity of organisms freely floating in water as 1. Then the volume corresponds approximately to the fresh weight. Then $10^6\ \mu^3 = 1\ \mu g$. If one multiplies the number of cells of individual species with the specific volume and adds up all the values, one obtains the total volume of the biomass of the algae or their

TABLE 5. IMPORTANT METHODS FOR THE INVESTIGATION OF PLANKTON

Name of the method	First used by	Apparatus required	Critical appraisal	Literature
Capture with a net	In the sea: J. Müller In fresh water: Apstein, Hensen	Plankton net	Exclusively useful for qualitative work on the meso- and micro-plankton; nannoplankton and ultraplankton are not captured	Hensen, 1887 Volk, 1901
Quantitative methods				
Chambers	Kolkwitz	Flat chambers of precise volumes (0·5, 1, 2 ml)	Only advisable for nanno-plankton and ultra-plankton with higher plankton densities	Kolkwitz, 1907, 1911
Sedimentation methods	Volk	No special apparatus needed	Degree of the plankton density can be adjusted to the conditions at the time. Rather cumbersome to operate	Volk, 1901 Glenk, 1954
Tubular chamber method	Utermöhl	Inverted microscope; special tubular chambers	Advantages of the chambers and sedimentation methods are combined; the best limnological method	Utermöhl, 1925, 1958
Centrifuge method	Lohmann	Electric centrifuge (a hand centrifuge is not suitable)	Possible to investigate the whole plankton in a living state	Lohmann, 1908
Membrane filtration	Kolkwitz	Membrane filtration apparatus	The whole plankton is captured but delicate organisms are deformed; convenient to operate by hand	Kolkwitz, 1924 Schmitz, 1950
Chlorophyll estimation	Harvey	Membrane filtration apparatus and photometer	Very economical of time; only the total plankton is captured; no individual values; suitable only for problems of production biology	Harvey, 1934 Handke, 1941 Gessner, 1942 1943, 1949
Estimation of the photosynthesis with C_{14}	In the sea: Steemann Nielsen In fresh water: Rodhe	Geiger–Müller counter	Rather troublesome; suitable only for determining the primary production	Steemann Nielsen, 1952 Strickland, 1960 Findenegg, 1964 Rodhe, 1958

total weight. It is troublesome to carry out this estimation but it is, on the whole, the only practicable way. Heinrich (1934), Ruttner (1938, 1952), and recently Nauwerck (1963), have obtained important results by this method. Unfortunately the size of the algae varies very greatly from lake to lake, so that the values obtained by the authors mentioned for the individual species cannot be generally used (Table 5).

Chlorophyll methods

Often chlorophyll extracts from quantitative phytoplankton samples are prepared for large comparative series, and the amount of chlorophyll per litre is determined colorimetrically. These results give only the total content of chlorophyll but in no way provide a criterion of the biomass and give no information at all about the relationships between the autotrophic, mixotrophic, and heterotrophic species of algae. For the precise method of determination, see p. 137.

For the estimation of the assimilation capacity of the algae (primary producers) in a lake, see pp. 137 ff. On p. 64 the development of phytoplankton methods is repeated with the aid of a summary by Glenk (1962).

Statement of the results

The results obtained with zooplankton and phytoplankton can be stated in the same way.

Vertical distribution

For statements of the vertical distribution the so-called spherical (Kugel) curves are used, that is to say the number of plankton organisms found at each depth in the lake is not given in absolute numbers per litre but as the cube root of the individual number divided by 4:

$$\sqrt[3]{\left(\frac{n}{4\cdot 19}\right)}.$$

This corresponds to the radius of a sphere (Lohmann, 1908). Cubic curves can also be constructed, in which the $\sqrt[3]{n}$ is directly shown, but in doing this one should always note whether spherical or cubic curves are meant (Thomasson, 1963). The advantage of this kind of graphic statement is that the large variations in plankton content are levelled out. Thus for 100 plankton organisms per litre $\sqrt[3]{n/4}$ gives the number 3, and for 100,000 per litre only 30. A further advantage is that the "linear population density" can be read off these "cube root curves" (Berger). The respective ordinates correspond directly to the number of organisms found in a given linear distance of water, a fact to which Berger (1965) has recently again called attention. Spherical and cubic curves are especially found in hydrobiology where diagrams of the vertical distribution are constructed (Fig. 44).

Seasonal change of the vertical population density

Nauwerck (1963) has found a very impressive method of demonstrating the development of populations in a lake in the course of the year, as an example of which the diatoms may here be shown (Fig. 45, in the original shades of grey). The abscissa shows the months and the ordinates the depths in metres, and the total volume of cells is given above. In every investigation the number of cells found per litre for the individual depths is included in

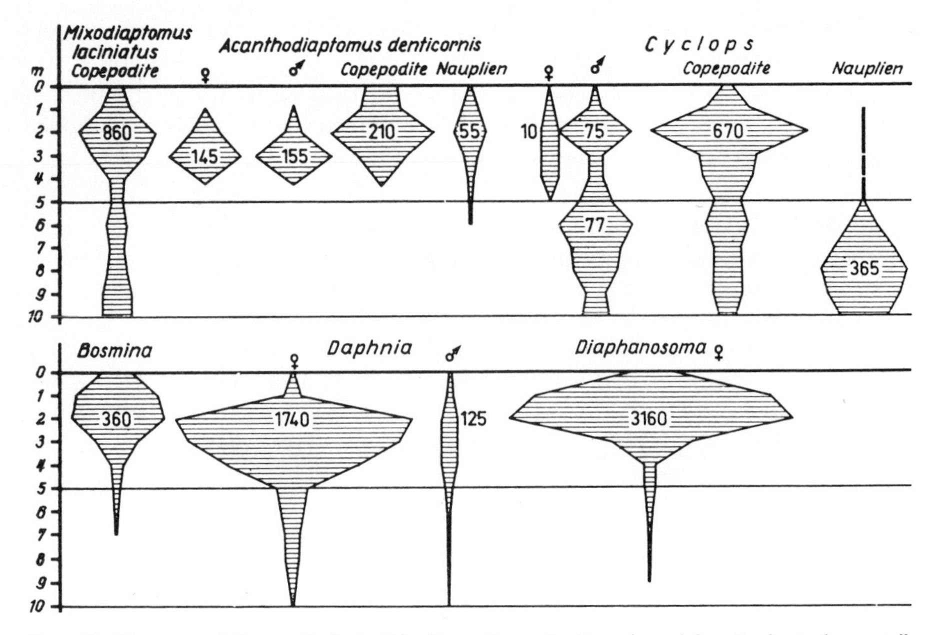

FIG. 44. Diagram of the vertical stratification of zooplankton in a lake: "spherical curves". (From Eichhorn, 1956.)

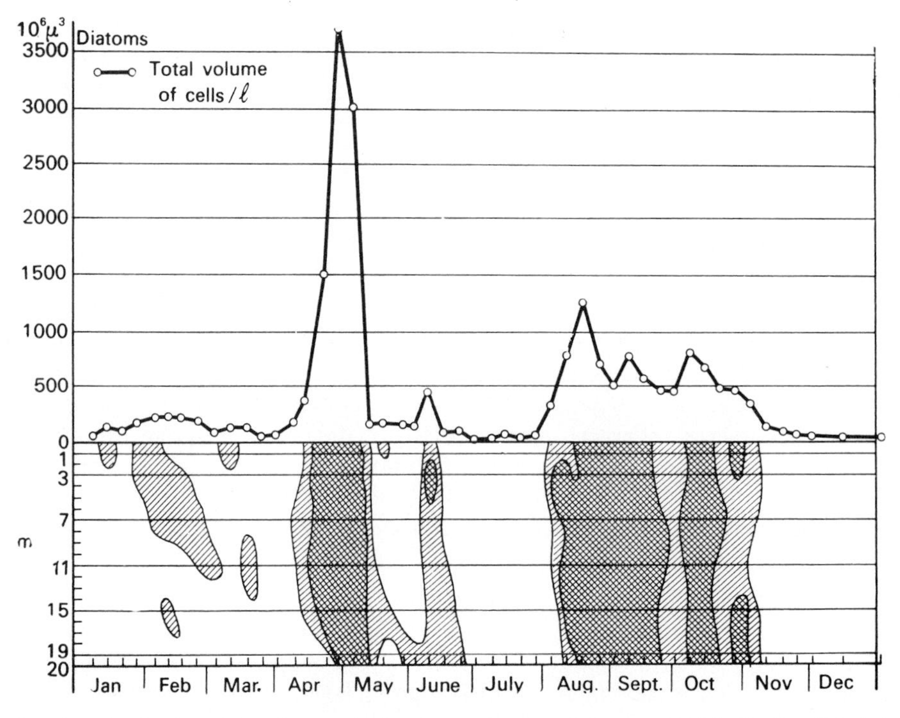

FIG. 45. Distribution of the diatoms in the various months of the year 1957 in the Swedish Lake Erken. (From Nauwerck, 1963, redrawn.)

the diagram and in addition the main feature of the quantitative development are shown by different shades of grey. The diagram thus combines the biomasses calculated from the volumes.

B. INVESTIGATION OF THE LIVING COMMUNITY ON THE SURFACE FILM (NEUSTON)

By the term "neuston" we understand the community of organisms which utilize the stability of the surface film. They are chiefly immobile algae which settle on the surface film and either project below into the water (hyponeuston) or into the atmosphere above (epineuston) (Fig. 46).

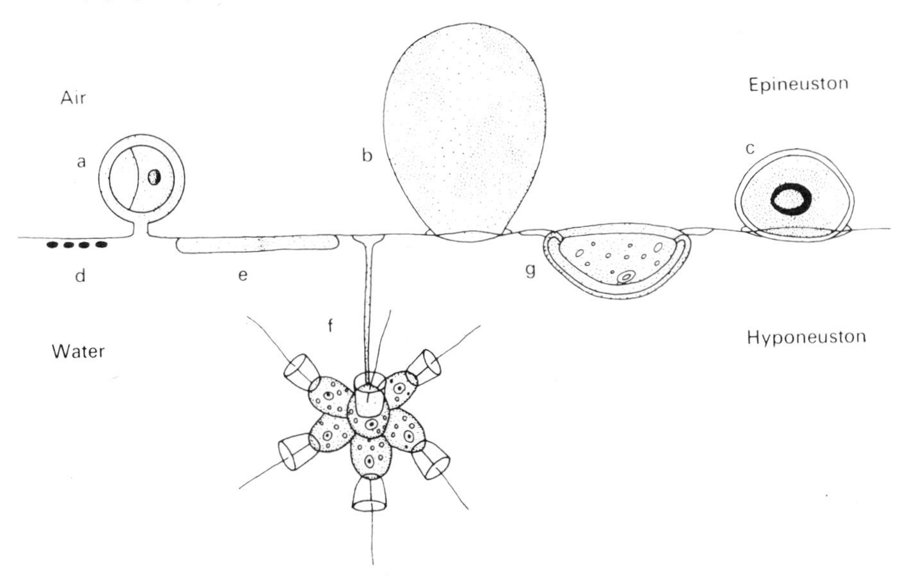

FIG. 46. Community of the surface film (neuston). (a) *Chromophyton rosanoffi* (Chrysophyceae), (b) *Botrydiopsis arhiza* (Heterocontae), (c) *Nautococcus emersus* (Protococcaceae), (d) *Lampropedia halina* (Coccaceae), (e) *Navicula* sp. (Diatomaceae), (f) *Codonosiga botrytis*, (Craspedomonapaceae), (g) *Arcella* sp. (Rhizopoda). (From Ruttner.)

In addition some animals can be included in the neuston—at any rate, temporarily. Among these are the cladoceran *Scapholeberis* and the ostracod *Notodromas monacha*, both of which apply their ventral surfaces to the surface film and glide along it and feed on the neuston. Among the aerial insects the skaters (Gerridae) and the water-striders (Hydrometridae) (*Velia*) live on the water surface. *Velia* utilizes indeed a specific reaction of the water surface for escaping upwards in that the insect squirts a fluid through its backwardly directed proboscis on to the surface of the water which lowers the surface tension and the expansion of the surface propels the animal forwards with remarkable rapidity. The same behaviour is known in the small beetle *Stenus*.

The micro-organisms attached to the surface film can be investigated in a very simple way. An ordinary cover-glass such as is used in microscopical technique is placed on the water in such a way that its whole surface is simultaneously applied to the water surface. This is best done with a pair of forceps which hold the cover-glass perpendicularly from above and lower it momentarily on to the surface. If one previously examines the water surface with a lens, one can find especially suitable places. The cover-glass, with the layer on the underside, is then put in a small block glass dish or on a Kolkwitz plankton counting chamber (see Phytoplankton, p. 59). If the specimen is to be fixed, one puts into the small dish or the counting chamber some 1% osmium tetroxide (OsO_4) aqueous solution.

This simple method derives from Naumann (1915) and was especially developed by Rylov (1925). Rylov (1926) used, for taking neuston, as well as a cover-glass, a small wire ring with which he removed the surface film from the surface of the water.

The organisms adhering to the cover-glass are directly accessible to investigation. In order to estimate them quantitatively also, Naumann photographed individual sectors, counted the organisms on the prints, and converted the result to the whole surface. Rylov (1925) made, with a camera lucida, accurate surface drawings on paper of individual visual fields and then counted them. Naturally in doing this one must always work with the same magnification.

CHAPTER 3

INVESTIGATION OF THE BOTTOM ZONE OF STANDING WATER (LITTORAL AND PROFUNDAL)

I. INTRODUCTION

Organisms of the bottom zone, the benthal zone, are not only exposed to those environmental factors such as light, temperature, and pressure which change, as they do in the pelagial zone, at different depths, but they also come into immediate relationship with the various substrates. The bottom zone of standing water is therefore divided up biologically more definitely than the open water zone is, and the methods of studying it are correspondingly more varied.

If one follows the gradient of the bottom out towards the lake centre, the depth of water increases only gradually from the shore and then suddenly increases at the slope down to the depths (Fig. 47). The division between the shore (littoral) area and the deep (profundal) area of a lake is based on structural criteria, but it is nevertheless very variable (on this, see especially Lenz, 1928, and Naumann, *Limnological Terminology*, 1931). From the point of view of the dynamics of metabolism of modern limnology it may, however, be best to designate, as Ruttner (1962) does, the whole of the profile of the bottom available to the photo-autotrophic plants as the littoral zone and the bottom zone below it as the profundal zone.

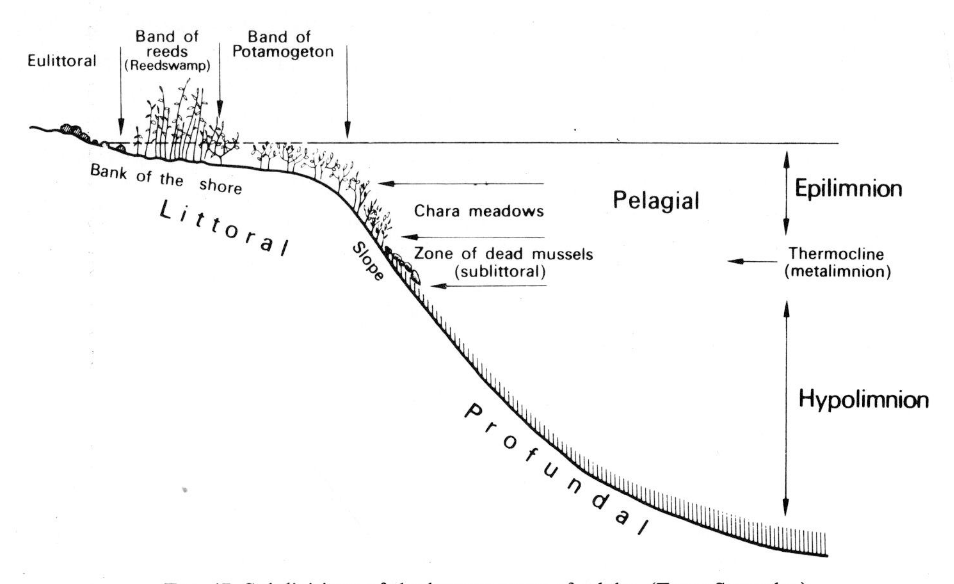

FIG. 47. Subdivisions of the bottom zone of a lake. (From Strenzke.)

The advantage of this conception is that the littoral zone can be incorporated in the trophogenic zone together with the corresponding layer of the pelagial zone because the positive balance of assimilation by the photoautotrophic plants, i.e the preponderance of the primary production, is characteristic of both these zones. The vertical extent of the littoral zone and of the trophogenic zone is altered in a similar sense according to the optical characteristics of the water.

It must be borne in mind that the distribution in depth of the phanerogamic plants is limited not only by the light but also by the pressure (Gessner, 1952; Golubič, 1963); but this does not hold for the algae. The horizontal extent of the littoral zone also depends on the structure of the shore. When the shore descends steeply the littoral zone is perhaps only a few metres broad or it has no horizontal extension at all (perpendicular walls of rock). When the shore is flat it may extend far into the lake. Shallow pools and small pieces of water usually have no true profundal zone at all. In such instances the primary production of the littoral plants plays a paramount part in the metabolism of the water. In the deeper lakes, on the other hand, the primary production of the littoral zone is relatively small in comparison with that of the trophogenic zone, and the chief part is played by the pelagial zone (cf. section on Phytoplankton, pp. 52 ff., and study of production, pp. 134 ff).

The organisms of the profundal zone live on the production of the trophogenic zone. In this zone only bacteria and some algae succeed in feeding autotrophically by chemosynthesis, and only a few diatoms adapted to extremely weak light are probably photoautotrophic. The organic production of the profundal zone thus proceeds by the disintegration and consumption of the primary and subsequent production of the pelagial and littoral zones. We can therefore place the profundal zone, in spite of its high secondary production, as the tropholytic zone of the benthal zone in contrast to the littoral zone and compare it with the tropholytic zone of the pelagial zone.

It may be pointed out here that the zones epi-, meta-, and hypolimnion characterized by their temperature conditions have basically nothing to do with the divisions into trophogenic and tropholytic zones; the epilimnion may coincide with the trophogenic zone, but need not do so in any way.

In the further subdivisions of the bottom zone we again follow Ruttner. The uppermost of the littoral zone lies in the region of the annual variations in the level of the water and of the shocks of the waves; this is the eulittoral zone. Typical environments are the interstitial ground water on the shore to which extensive attention has been given in the last 25 years, and the region of breakers (surf) in lakes. Those parts of the shore which are always under water belong to the true littoral zone (but are still called sublittoral by Ruttner). It is best to make subdivisions according to the bands of vegetation which succeed one another. We find in succession:

1. The plants on the bank of the shore above the water (band of reeds).
2. The potamogeton zone and the band of plants floating with leaves at the edge of the bank of the shore and of the slope downwards.
3. The region of the underwater meadows on the uppermost part of the slope down (see Fig. 47). In extremely transparent lakes this may be succeeded by a zone with clumps of green algae (chiefly *Spirogyra*) (e.g. in Lake Vrana in Jugoslavia; cf. Golubič).

The profundal zone is further subdivided according to the various kinds of sediment on it; cf. Lenz (1928), the very detailed *Limnological Terminology* of Naumann (1931), and Züllig (1956).

II. INVESTIGATION OF THE EULITTORAL ZONE

The eulittoral zone comprises the outermost zone of the shore of a lake in the region of the direct shock of the waves and of variation in the level of the water. Its extent depends on local conditions such as the steepness of the shore and the extent of the variation of the level of the water. Investigation of it is carried out without using a boat because of the strong force of the waves and the shallow depth of the water.

1. The Interstitial Ground Water of the Shore

The pore spaces in the sand and shingle of the eulittoral region are filled more or less with water. In them live animals which have very often a quite characteristic form, which are found also in similar environments on the coasts of seas (mesopsammal) and on the banks of running water (hyporheal, cf. p. 118). Algae are also found in the uppermost layers of these deposits. The work of Pennak (1940, 1950), Neel (1948), Ruttner-Kolisko (1953, 1954, 1962), Ponyi (1960), and others has provided us with a great deal of knowledge about the conditions of life in these pore spaces and about their populations. When samples are taken one must bear in mind that differences in the distribution of the organisms occur in both a horizontal and vertical direction. The hygropsammon, over which the waves break and of which the pore spaces are completely saturated with water, is differentiated from the eupsammon, which is not reached by the waves. Two methods of study are available.

Diggings in the immediate neighbourhood of the shore

A hole is dug in the hygropsammon with a small spade as far down as the underground water. The water collected from the pore spaces is drained out several times and filtered through a hand net (Fig. 48), made with gauze No. 25. This also holds back the detritus

Fig. 48. Simple hand net for collecting organisms on the shore of a lake, in small collections of water, and in mountain streams.

which often densely fills the pore spaces and on which the organisms feed. The sample of detritus thus obtained is taken quickly to the laboratory without any addition of fixative to it and the organisms are counted there.

Before the counting, the volume of the whole sample is determined in a graduated cylinder. A few millilitres of the mud sample, first diluted with a little water, are distributed in a thin layer on a Petri dish, the bottom of which has been divided by parallel lines. Through the binocular, everything present within a band is counted and the whole Petri dish is covered in this way. Five to ten per cent of the whole sample, depending on the amount of organisms present, is counted in this way, and, finally, the number of organisms in the whole catch calculated.

By means of diggings of this kind at various distances from the shore and at various depths one can determine the distribution of the organisms in the sediment (Pennak, 1940; Ponyi, 1960). In this manner Ponyi was able to find, in Lake Balaton in Hungary, that the plankton Crustacea *Daphnia hyalina cucullata*, *Scapholeberis kingi*, and *Mesocyclops leuckarti* are washed by the waves into the interstitial spaces and can maintain themselves in these for some time.

General method of taking a sample

This is done by means of a small tube which is driven perpendicularly into the sand. The column of sand thus obtained can be carefully divided into discs much as one pleases, but scarcely less than 1 cm thick, which are then examined separately under the microscope (method of Neel, 1948).

By this method Ruttner-Kolisko studied the vertical distribution of organisms on the sandy shore of the Swedish Lake Erken. To collect the organisms the sand sample is thrust against a solid substrate until there is sufficient supernatant pore water. One cubic centimetre is pipetted off this and the organisms in it are quantitatively counted. The water content of each of the samples is estimated by the difference between the weights of the wet and dried samples and from this the total content of organisms is determined. The organisms are stated per volume of sand.

A still better method of collecting organisms from sand samples was devised by Remane; a weak solution of formalin is added to the sample which, after shaking, is decanted from the sand. The formalin narcotizes the organisms and prevents them from attaching themselves to the sand grains and from sinking with these to the bottom. Thus they remain "in solution" and can be collected after repeated washings and decanting.

The population density often reaches 1000 individuals per cc of sand, the diatoms, flagellates, ciliates, and rotifers being especially numerous and also Crustacea, tardigrades, worms, and Gastrotricha, but mites are only rarely present. As a general principle the temperature in the sand and in the neighbouring surface water should be taken and also an analysis of the particle sizes of the sand should be carried out. The works of Pennak and Ruttner-Kolisko especially contain hydrographical and chemical information.

2. Investigation of the Surf Zone on the Shore

The surf zone is defined, in contrast to all the other littoral zones, by very characteristic conditions of life. In the first place it is the region in which the movement of water in the lake is most pronounced. In it extreme annual and daily temperature variations occur. Further, the oxygen content of the water is always high in comparison with that of other regions of the lake and especially with the deep zone. All these factors directly affect the animal and plant organisms and give to the biocoenosis within the lake a quite characteristic composition which in many respects resembles that found in running water: the movements of the water and the good oxygen supply are decisive conditions of life in both these habitats.

The higher plants hardly maintain themselves here, and only the algae form more or less firm coverings on the stones. Many animals also attach themselves to the stones or seek between them protection from the vigorous movements of the water. Wesenberg-Lund (1908) and Ehrenberg (1957) have especially studied the animal colonization, and Kann (1941) has made a very thorough study of the algae coverings of the stones.

Plants (epilithic algae)

The surf zone is either sandy or stony. Numerous epilithic algae live on the stones in the form of algal incrustations, and it is difficult to collect these quantitatively. Epilithic algae are also found on clumps of artificial stone and on concrete walls.

Kann has devised a rather satisfactory method for these. In its simplest form it is to scrape or lift off the algal growth with a knife. Many species, especially some blue–green algae, grow in a very thin layer on or in the substrate; and these species can be removed with a razor blade only in favourable instances. The method described by Douglas for running water (see pp. 107 ff.) may also be useful for this purpose.

Calcareous stones with thick and porous incrustations can be removed together with the algae on them. The incrustation is carefully decalcified with weak hydrochloric acid and vertical sections as thin as possible can be made with a razor blade; their algal content can then be investigated. For instances in which the algal incrustation cannot be readily removed from the stone, Golubič recommends sprinkling very weak hydrochloric acid on them or putting them completely in this. After some hours or even days the inctrustations can be lifted off and is then also simultaneously decalcified.† Instead of hydrochloric acid one can also use lactic acid, only this takes longer to act.

For additional quantitative methods see the section on running water, pp. 106 ff.

Animal population

Among the animals inhabiting the surf region we meet with many species also met with in the study of running water. The constant movement of the water in the surf region is much more critical for the animals in so far as the direction of the incoming and outgoing waves is constantly changing. In contrast to the conditions in running water, the temperature in the surf region greatly increases in the summer and especially undergoes great changes during the day. This never occurs to the same degree in running water. Thus in both these habitats, only those organisms live which are especially tolerant of the temperature. At any rate in the surf region of East Holstein lakes, only 72 out of 155 species of animals were typical of this habitat and 83 also lived in running waters in the plains and mountains.

The animals attach themselves to stones, and live between stones or in the sand. If the stones are exceptionally overgrown with mosses, they seek shelter in these growths. Otherwise they live in algal growths or attach themselves directly to the smooth stone. Because of these varied modes of life, it is difficult to collect the animals quantitatively from a given substrate.

The methods adapted for the study of this habitat are very simple. To make a qualitative exploration of the fauna one takes out a stone and collects the animals attached to it with forceps and holds under the stone a small hand net (see Fig. 48) which catches the organisms dislodged and also those which allow themselves to fall off the stone (as a flight reaction). The catch is then either immediately fixed or carried alive in a large collecting vessel. Best for this purpose are vacuum flasks, insulated against temperature effects, which can be bought in all sizes.

† I am grateful to Dr. Golubič for this verbal information.

In deeper water the stones can be taken out by hand or with a scraper on a handle (Fig. 49). The lower edge of this is a sharpened iron bar directed forwards with which the stones are loosened so that they can be collected in the bag of the net.

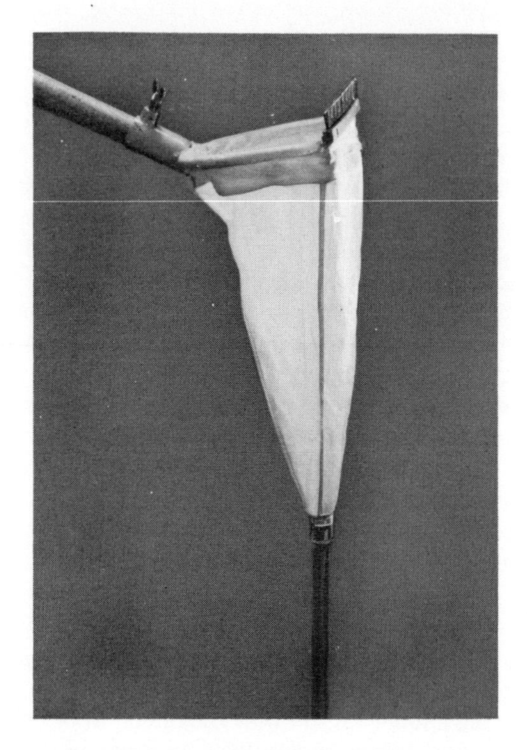

FIG. 49. A scraper on a handle, provided in front with a brush for brushing off objects under water. For use among heaps of stones it has, instead of the brush, a sharp edge made of iron.

Bottom grabs, dredges, mud-samplers, and similar instruments are not suitable for quantitative work because they do not penetrate deeply into hard substrates, or do not penetrate deeply enough. It is best to separate off a certain part of the bottom, about 1 m^2, by means of an iron frame the sides of which are 1 m long, and to collect the stones inside it; in deeper water one uses a box with high sides and a smaller surface. Because the animals live on the surface of the stones it is important to estimate this surface. A direct method of doing this by estimating the surface of a stone by planimetry of all surfaces is possible, but much too troublesome, and it cannot be done in a reasonable time. Schräder (1932) has therefore recommended that the greatest area (the largest projection from above) should be taken as the colonization area. This greatest surface area would be determined for each of the stones lying in the delimited area, and all the individual surface areas would be added together, and the total surface area thus obtained would be related to the number of organisms obtained. Possibly more comparable results would be achieved if one reckoned the stone surfaces as the surface of a sphere of the same volume. To do this the volume of a stone is estimated by determining its displacement of water, and from $\frac{4}{3}\pi r^3$ the radius of a sphere of the same content is calculated. The surface can then be determined from $4\pi r^2$. To do this estimation for every stone is very troublesome; Ehrenberg has therefore estimated the surface of spheres corresponding to small, medium-sized, and large stones and has estimated by this method the sphere surfaces of the stones examined.

The animal population is calculated for 1 m^2 of stone surface.

It is also difficult to use bottom grabs on sandy bottoms. Quantitative samples can be taken with tubes open at both ends and driven into the sand. Naturally this can only be done in shallow water and provides only very limited sand samples which are evaluated by the method used in the study of interstitial ground water of the shore.

III. INVESTIGATION OF THE TRUE LITTORAL ZONE (INFRALITTORAL)

In the methodology of study of this zone three bands of plants must be differentiated which succeed one another in the direction of the middle of a lake, and each of them provides different conditions of life and differing colonization. These are; the reedswamp, with especially *Scirpus lacustris*, *Phragmites communis*, and *Typha;* the zone of *Potamogeton* species and plants with floating leaves, *Nymphaea*, *Nuphar*, *Hydrocharis*, and others; and finally the zone of submerged meadows with *Myriophyllum*, *Ceratophyllum*, *Elodea*, and especially the Characeae, extending to the deepest depths. It has already been mentioned that the pressure of the water delimits the distribution in depth of the phanerogams, but not that of the Characeae.

1. Reedswamp Zone (Upper Infralittoral)

This zone constitutes an extensive habitat that is separated from the open surface of the lake. It consists essentially of the common reed *Phragmites communis*, the bulrush *Typha* together with several other species, and the rush *Scirpus lacustris*. The surf loses itself to a greater or lesser degree in the forest of reed stalks according to the density of the belt of plants. There is, therefore, no attachment of the organisms in this still-water region as there is in the surf zone. But lack of exchange with the open lake also makes aeration difficult, so that the oxygen supply is not favourable as it is in the surf zone; it varies very much in different parts of the zone of reeds, and this must be borne in mind when accurate hydrobiological investigations are being done.

The organisms in this belt of plants live either between the stalks—partly even planktonic—or they live on the bottom or on the stalks of the plants themselves in the *Aufwuchs* which consists of both plants and animals. This growth is called the periphyton. A complete investigation must include all these three biotopes and they must be studied with different methods.

The organisms in the open water are captured with hand nets, with the scraper, and with plankton nets of various widths of mesh mounted on poles. Among the plankton organisms found here are especially algae; among the animals, insect larvae, rotifers, and small Crustacea, the two latter being planktonic. Only the planktonic species can be taken quantitatively, i.e. in relation to a volume of water (of course with samplers), but not in the water-bottles discussed for plankton but by means of simple beakers or pails. The contents of these samples are then filtered through a coarse net which filters out the larger animals and then through a No. 25 gauze which filters out the plant macroplankton and the animal microplankton; the water that is collected afterwards contains nannoplankton. All of these three fractions must be evaluated separately (for this, see the section on plankton, pp. 33 ff).

Among the organisms on the bottom, animals preponderate. They are best collected with a scraper on a handle with which mud is taken from the bottom and sieved. Naturally nothing can be done in the belt of reeds with dredges and bottom grabs.

The periphyton *(Aufwuchs)* consists of a more or less dense coating chiefly consisting of filamentous algae and diatoms to which particles of mud adhere so that an almost muddy coating is formed, which itself harbours an animal world rich in species and individuals.

If one wishes to understand correctly this living community quantitatively, one must remove the plant growth from its substrate and consider also all the animals living in it. It is scarcely possible to achieve this completely. Ponyi (1962) has shown, by his beautiful researches, that the lower Crustacea are bound to the plant growth only very loosely and that light shaking detaches them from this. This does not apply to the nematodes, bdelloid rotifers, Testacea, and all the other firmly attached species, so that there is a risk that the method of collecting may take only a selection of the living community.

To collect the periphyton quantitatively, Meschkat (1934) cut the plant stalks close to the bottom with a long-handled sickle and removed them or he first cut off their heads and put a long glass tube over the stalk before removing it. Finally he cut the stalk into 20 cm lengths, numbered them consecutively, and transported them in suitable glass tubes. The pieces of stalk must be studied separately because differences are to be expected in the vertical succession both in the plant growths and also in the animal populations.

By this treatment the Entomostraca are, as Ponyi has shown, almost completely lost. In reed stalks in Lake Balaton the number of species of them sank from 21 to 4–5. Meschkat's method is thus not quantitative for Entomostraca, but it gives good results in the study of the vertical distribution of the growth on the reed stalks.

Schönborn (1962) collected the Testacea (amoebae with shells) in the growth by scraping off the growth on stalks of *Typha* and *Phragmites* for a length of 25 cm and put it in a vessel with water. In this a length of 25 cm corresponds to a surface of 225 cm^2 so that it was possible to make a direct comparison with the surface sampled by the Ekman–Birge bottom grab (cf. p. 85).

Schönborn squeezed out the algal growth in the laboratory and distributed the detritus evenly in a known amount of water. He then counted 50 cm^3 of this sample and calculated from this the total content of Testacea in the sample. The plant and animal populations were thus related to the surface of the stalks. But it must be remembered that a much greater substrate surface in periphyton is available for the animals.

Meschkat scraped off the growth first when in the laboratory and examined it under a microscope without previous fixation so that the larger animals were quickly counted. The water in the tubes used for transport was treated with formalin and the organisms attached to the glass walls were brushed off. After sedimentation for 24 hr the deposit in the tubes used for transport was removed and counted together with the sample fixed with formalin in a Petri dish with a chequered bottom.

It is best to fix and preserve samples of algae that are not to be studied immediately with a 2% solution of formalin and to preserve them. The amount of plant growth on the pieces of reed studied may be also stated quantitatively by estimation of the volume or weight.

In statements of the results of investigation of the periphyton, the following data are important: the specific composition and amount of the algal growth and the number of animals in relation to the surface area of the substrate; the specific composition of the animal communities; the vertical zonation depending on the light gradients and other factors. This vertical distribution has been especially studied by Thomasson (1925), v. Cholnoky (1927), Ivlev (1929), Meschkat (1934), and Schönborn (1962). For work on the *Phragmites* belt see also Gessner-Liersch (1950), and for the fauna of the algal growth, Meuche (1939).

2. Zone of Floating Leaves and the *Potamogeton* Zone (Middle infralittoral)

These zones (Fig. 50) connect up with the zone of reeds and rushes but are found only where the movement of the waves is not too strong, e.g. in bays sheltered from the wind. The zone of floating leaves plays an important part in the lives of many animals, especially for laying of eggs and nutrition (see Fig. 1). The zone of floating leaves is formed especially of growths of water lilies and species of *Potamogeton* with floating leaves, such as *P. natans* and *P. alpinus* and the water knot-grass, *Polygonum amphibium* L., and the frog-bit *Hydrocharis morsus ranae* L. Th. v. Bülow (Dissertation, Univ. Kiel, 1951), Krasowska and Mikulski (1960), and Müller-Liebenau (1956) have been chiefly concerned with the animal colonization of this zone.

Study of the free water of this zone is carried out by the methods used to study the pelagial area, and it need not be further described here.

For the study of the bottom of this plant zone, on the other hand, some apparatus is available which must be described in further detail, namely dredges for taking qualitative and bottom-grabs for taking quantitative samples. Because the bottom grabs are the most important aids in the study of the characteristic deep fauna of the water-bodies, they are described in detail in the section on the profundal zone (pp. 85 ff.).

There are various kinds of dredges. The one most often used is the simple net dredge of Steinmann (Fig. 51). It consists of a heavy triangular iron frame to which a net bag of coarse sand gauze (cf. p. 37) is attached and over this—to protect it—a bag of coarse linen. At the angles of the metal frame there are holes or rings for the attachment of the draw

Fig. 50. Floating leaves of *Potamogeton natans* in a small piece of water.

FIG. 51. Steinmann's dredge for investigation of the bottom.

lines. Lundbeck, Kolkwitz, Fittkau, and others have also constructed dredges with quadrangular frames (Fig. 52), all of which operate according to the following principle: the dredges are pulled along on a long line; because of their weight they readily press into the bottom and thus collect the mud from the uppermost layer of this, but without sieving it, as a net would do. There is therefore no object in pulling a dredge along for longer than until its bag is filled: this can easily be roughly estimated from the length of the rim of the frame. In this connection reference may be made to the closing dredge of Riedl, which is described in more detail on pp. 94 ff.

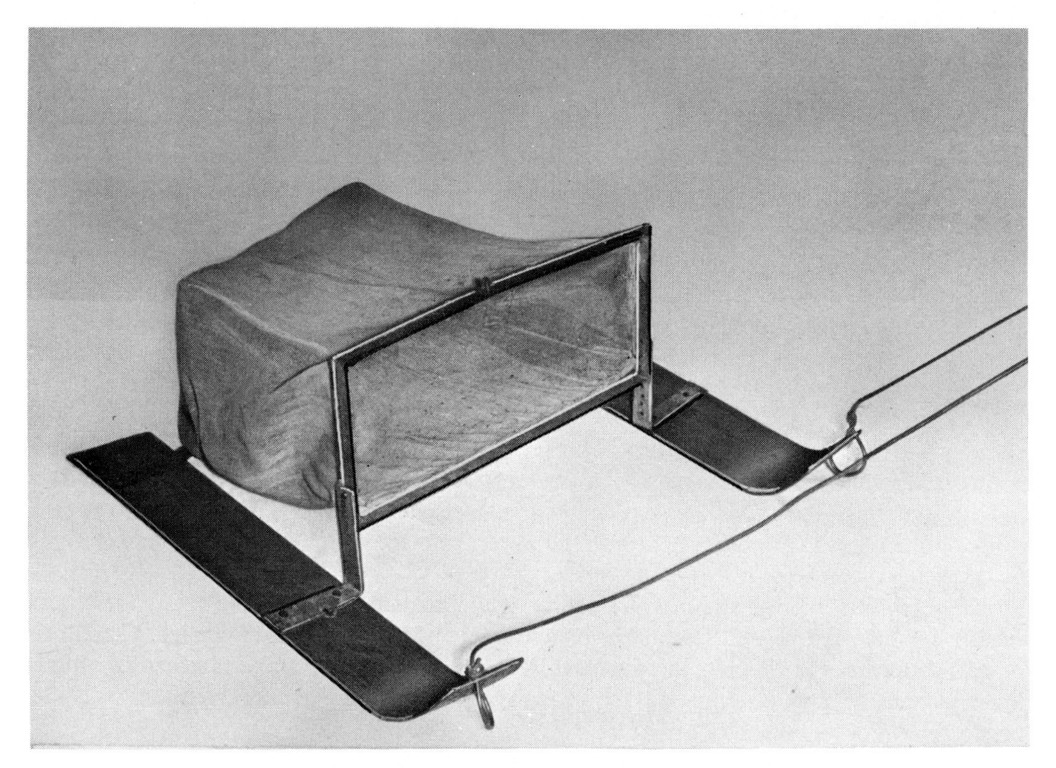

FIG. 52. Rectangular dredge.

After the tow is finished the dredge is pulled up and its contents are, in portions, freed from detritus by means of a net made of sand gauze by dangling this in water; this must naturally be done with care so that no animals get into the sample from above. Metal sieves, e.g. the mud sieve of Kolkwitz, are not suitable because forms with soft skins, especially Turbellaria and other worms, are readily broken up (from the method of sieving, see pp. 104 ff.). The residue in the net is put, in portions, into a white bowl with a little water and the animals are picked out with pipettes or forceps. According to the plans of the investigation, the organisms are fixed on the spot (see Appendix II) or they are transported alive in wide collecting vessels.

The results of the catches made with the dredges mentioned are not quantitative, i.e. the catches cannot be accurately related to a bottom surface area from which they were collected. Dredges have been constructed for making quantitative catches, but these cannot be used in the plant belt of the littoral zone; they are described in the section on the profundal zone together with bottom grabs.

The colonization of the plants themselves

The plants are collected up with rakes, iron hooks, fishing lines, and dredges. The plants with floating leaves are also encrusted with an algal growth, but this is not so thick as it is in the belt of reeds. In the study of this growth a distinction must be strictly made between the stalks and the floating leaves. The floating leaves are chiefly populated on their undersides, especially by leeches, oligochates, molluscs, and insect larvae, which often cut pieces of a characteristic shape out of the leaves or eat out tunnels in the tissue of the leaves between their upper and lower surfaces (leaf miners). A typical miner in the floating leaves of *Potamogeton natans* is the larva of the chironomid *Eucricotopus brevipalpis*.

3. The Submerged Plants (Underwater Meadows) (Under Infralittoral)

The completely submerged plants populate the littoral region furthest from the shore as far as a depth at which the weak light still suffices for a positive assimilation balance. This depth is reached only by the moss *Fontinalis*, the Characeae and filamentous algae *(Cladophora, Spirogyra)*, which form the normal underwater meadows of the lowest littoral region.

In this zone the gelatinous spheres formed by the algae *Nostoc* and *Aphanothece* or, when these are coloured a clear green by the infusorian, *Ophrydium*, are especially remarkable.

Among the higher aquatic plants the most important species are *Ranunculus*, *Myriophyllum*, *Ceratophyllum*, and various species of *Potamogeton* and *Elodea*. For charting the plant community from the boat an instrument for seeing underwater is useful. This "water-viewer" is box-like, cylindrical, or shaped like a truncated cone made of wood or metal with two lateral handles, its lower end being sealed with a glass plate, whilst its upper end allows free vision into it. This instrument is put into the water from the boat, and with it the plants can be conveniently inspected.

The organisms in the free water are collected and evaluated as usual with plankton nets or they are quantitatively estimated by catches made with water-bottles or pumps. The quota of pelagic forms is in this region of the littoral zone far from the shore already very high, and the zone of free water above the growths of *Chara* belongs wholly to the pelagial zone.

The organisms on the bottom, among which the plants are also to be included, are collected up by dredges and bottom grabs together with the bottom deposits, and are further examined as described on pp. 85 ff. For the special purpose of securing plants in the greater depths, rolls of barbed wire and iron rakes with spikes on both sides have been well proven. They are dragged along on long lines, whilst the plant grab constructed by Reith is let down perpendicularly and functions like a bottom grab. Together with the plant, only those organisms are brought up which are attached relatively firmly to them, i.e. usually the organisms growing on them; all the others are lost during the passage up. Garnet and Hunt (1965) use a cylinder made of sheet steel with a basal surface of 1/30 m^2 which is placed from above over the plants. A round, accurately fitting steel plate, the borders of which are ground sharp, is so rotated by means of a firmly mounted rod that the cylinder is closed below. In this way the plants are cut through more or less closely to the bottom and are brought up with the cylinder. Parts of the cylinder casing and also the bottom plate are covered with wire gauze so that the water can easily run out. This tubular collector is especially good for use in vegetation growing upright on level sandy or muddy bottoms.

The growth on these plants is also very rich. It consists chiefly of green algae and diatoms and harbours a characteristic fauna of nematodes and Testacea. For the reasons given above it is difficult to assess this community quantitatively.

The population of the zone of submerged plants is extraordinarily rich. Numerous species of animals live in the tangle of plants and use them for building shelters (houses), as sources of food and during their reproduction. In the study of these animal communities, therefore, attention should be paid to the fact that many species of animals make, in the littoral region, as happens also in the plankton, daily vertical migrations and at night move towards the surface (Mundie, 1959). Pieczyńska (1960) noted this among the aquatic water mites, and Szlauer (1963) studied it among Crustacea with small funnel-traps.

4. Semi-experimental Methods for the Study of the Periphyton (Aufwuchs)

Already Hentschel used artificial substrates, made of slate, wood, and celluloid, to study in more detail the development and composition of the periphyton. These substrates, usually plates, were exposed in various positions on the surface of the water and at various depths, and the growths on them were studied at various periods of time. Many changes in this method have since been introduced. Sládečková (1962) has very thoroughly summarized the methods hitherto used. Because of the importance of these methods, we must deal with them a little here, namely in the following sections: exposure of the material and the time of exposure, quantitative estimation by counting, and quantitative estimation by determination of the biomass. The methods used for running water will be described in detail on pp. 123 ff.

Exposure of the substrate

Because the colour of the substrate has no effect on the periphyton and the roughness of its surface has only a quantitative effect on it, it is best to use transparent substrates with a smooth surface made of celluloid, plastic, and especially glass. Microscopic slides serve especially well as glass plates because the growth on them can be directly examined under the microscope. Strips of celluloid have the advantage that they can be used in any lengths and provide a continuous horizontal or vertical profile. When the subsequent evaluation is being done they can be cut up into pieces of any size desired. Strips of celluloid can also be coiled around stakes or stones, and can be exposed in any place. In the reedswamp glass rods can also be put upright into the bottom.

Each worker has developed his own method of exposing microscope slides, and we can only give hints. Usually a horizontal and vertical position of the glass plates suffices. In very eutrophic and polluted waters plates set horizontally do not give good results because the intensive sedimentation results in their being very quickly covered with detritus, so that plates set vertically are better in such waters. In oligotrophic waters horizontal plates also give good results. In order to determine a profile of the periphyton colonization from the surface of the water to the bottom, several glass plates are exposed above one another. The distance between them near the surface should not be more than 20 cm and below this not more than 100 cm. The plates are fixed by means of lateral notches in the glass to cords hanging down vertically, which are themselves tied to a buoy and anchored to the bottom with a stone. It is also very simple to make a wooden frame which has along its long side grooves for the glass slides, as in a staining dish used for microscopic preparations. This frame and its slides can be exposed horizontally or vertically and it also can be attached to a buoy. Burbaneck and Allen (1947), Newcombe (1949, 1950), and Yount (1956) have obtained good results in this way.

Sládečková (1960, 1962) has somewhat modified Kusnetzow's very good method and has used it with very good results. The glass plates are attached to large cork stoppers in such a way that each cork carries four horizontal and four vertical glass slides (Fig. 53). Through a hole bored in the middle of the corks a rubber-insulated cable is passed. The corks with the glass plates are fastened at intervals of 1 m by knots in the line above and below the corks. The cable is attached to a buoy or wooden float though, of course, there is a risk that the uppermost corks exposed close under the surface may be shaded by the buoy. The cable is kept taut by means of a heavy stone on the bottom. The rubber insulation enables it to withstand exposure for more than one year. Pieczyńska (1964) also used

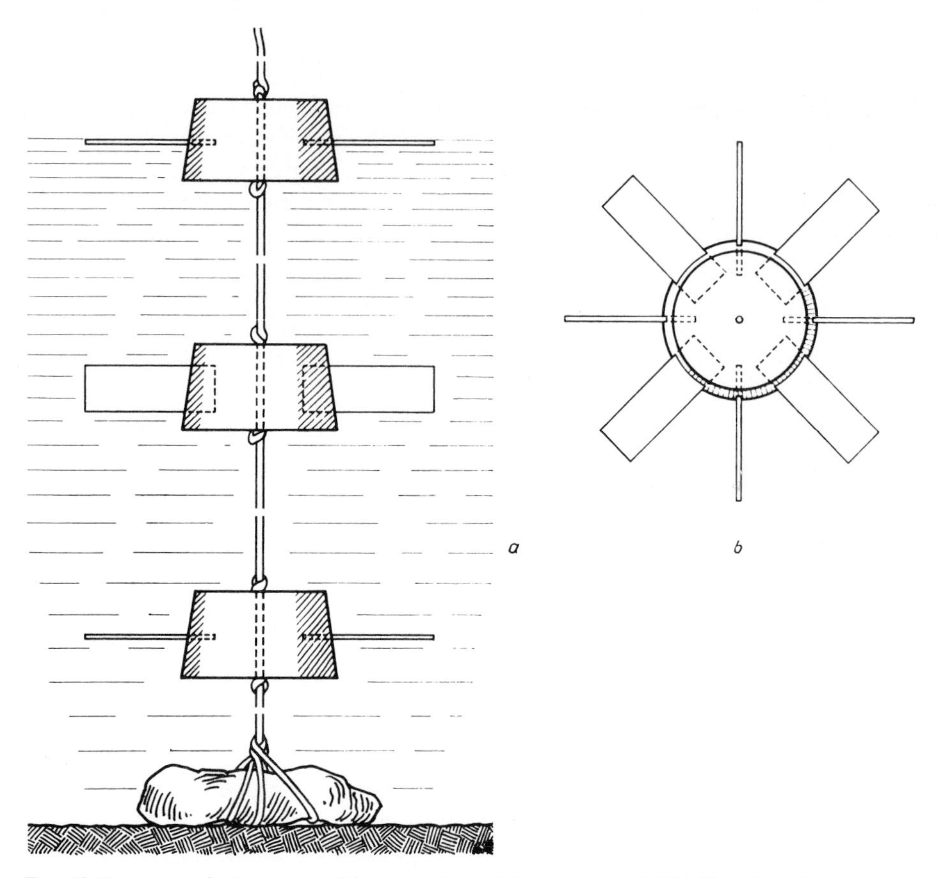

FIG. 53. Exposure of microscope slides according to Kusnetzow and Sládečková: (a) lateral view, (b) top view of the cork.

microscopic slides, but also larger glass plates held in a a wooden frame. The frame is perforated and is exposed, hanging in the water on a cord.

Collection and transport of the glass plates exposed

At intervals of one to several weeks one horizontal and one vertical glass slide is taken from each cork or, in other techniques of exposure, one proceeds in a corresponding manner. Each plate removed is replaced by a new one. The plates collected are transported individually to the laboratory in wide-necked flasks with lake water. Naturally each plate is carefully labelled with the exposure time, position, and water depth.

Determination of the growth by counting

Further work on the glass plates is done as soon as possible. Fixation with formalin or other reagents is unfavourable and is to be avoided. Because there is growth on both sides of the plates, the growth must be removed from one side. In plates exposed horizontally the growth on the upper and lower sides must be carefully differentiated from one another, but there is no difference of this kind in plates exposed vertically. The periphyton on the side cleaned up is quantitatively rinsed into a wide-necked flask. The cleaned and dried surface of the glass plate is laid on another slide, the middle surface of which is divided into areas of 1 cm^2; with a moderate magnification (about 100 times) 2 cm^2 of the middle of the slide are examined and all the organisms in this area are counted (for the technique see Zooplankton, pp. 51 ff.) with the exception of the free-living larger animals, which are counted separately from the total sample.

Each individual is counted in colonies of *Campanella umbellaria*, *Carchesium polypinum*, *Ophrydium sessile*, and other larger organisms; but colonies of very small flagellates, such as *Monadodendron*, *Condosiga*, etc., are counted as individuals. Also counted as colonies are the *Synedra* colonies divided into compartments, the filaments of blue-green algae and green algae and of *Melosira varians*, the branched filaments of the iron bacterium *Crenothrix fusca*, the gelatinous envelopes of *Tetraspora cylindrica*, the discs of *Coleochaete soluta* and others (Sládečková).

On slides exposed for a very long time in the upper layers of the water, the growth of diatoms and green algae is often so dense that the plates are almost or completely opaque, so that an accurate count is not possible. But one can adopt the expedient of removing the uppermost layer of the growth and distributing it in some water and counting it separately.

After the completion of the count, the growth is scraped off the glass plates and is preserved in 2% solution of formalin in wide-necked flasks for subsequent qualitative investigation.

The average count per square centimetre is taken as a basis for the calculation of the growth on each plate. The average value per square centimetre is then multiplied by the double surface of the glass plates, because both surfaces are overgrown, but those parts of the surfaces stuck into the corks or frames, etc., must be subtracted. The quantitative statement of results is made as the number of individuals per square centimetre.

Normally, larger animals attach themselves to the plates such as rotifers, worms, polyps, insect larvae and pupae, and also small Crustacea. These are counted separately from the whole plate.

Quantitative estimation of the periphyton by determination of the biomass

If the growth on the plates is so thick that the plate is not transparent, or with a largely opaque plate, a direct count under the microscope is not possible. Other quantitative estimations must be used, but these need not be described further here.

1. Fresh and dry weight estimation (e.g. Cooke, 1958 and Grzenda and Brehmer, 1960). Sládeček and Sládečková (1963) found that the dry weight of the growth estimated after heating to 105°C was about 8% of the fresh weight. If one determines, as an expedient, the dry weight in the air, this is on the average 8·8% of the fresh weight.
2. Determination of the fresh volume in graduated cylinders (e.g. Hentschel, 1916 and Sládečková, 1957).
3. Measurement of the photosynthesis and respiration in clear and dark bottles (see pp. 142 ff. by the method of Backhaus).
4. Chlorophyll determinations (see pp. 136 ff.).

Statement of the results

This cork-stopper method, like all others, provides data about the following relationships: qualitative and quantitative composition of the growth according to the period of exposure and the site of the exposure in relation to the water surface and depth of the water.

From these data conclusions can be drawn about the temporal structure of the biocoenosis under various conditions. Sládečková (1960) has constructed from them a very instructive space-time-diagram for individual species and groups of species (Fig. 54). There were very interesting differences in the animal colonization relative to the position and orientation of the plates.

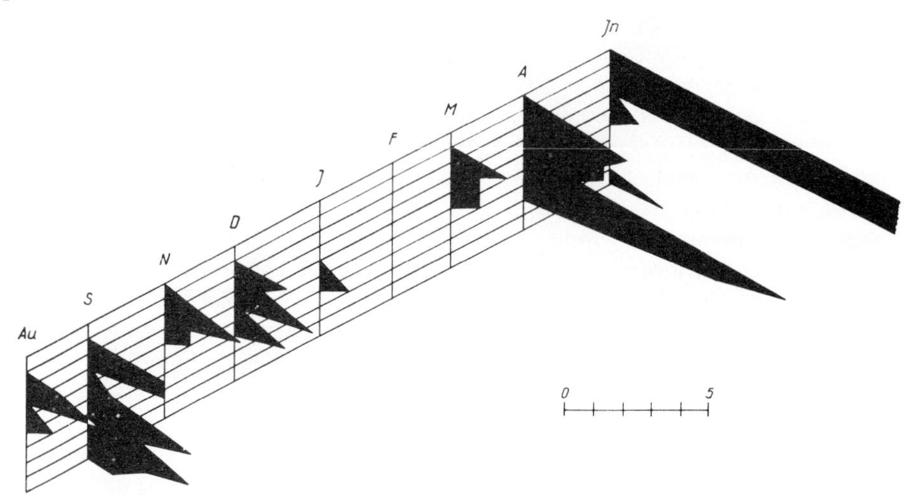

FIG. 54. Distribution of the diatom *Melosira varians* on glass plates exposed at various depths during various months. Abscissa: the months; ordinate: depth in metres. Scale: filaments/cm². (After Sládečková, 1960.)

IV. INVESTIGATION OF THE DEEP (PROFUNDAL) ZONE

The profundal region of lakes lacks photo-autotrophic primary production: the organisms in it live on the destruction and utilization of the organic substances synthesized in the trophogeme zone of the pelagial region, whilst the primary production of the benthal region is of only small importance; the littoral plants dying in the autumn are carried away by the wind and currents in the lake and then gradually sink to the bottom.

As was explained in the introduction, the oxygen conditions in the deep regions of a lake depend on the production of organic substances in the trophogenic zone and on the ratio of the trophogenic to the tropholytic zone. Thienemann (1918) has drawn attention to these variable oxygen conditions and also to the differences in the animal colonization conditioned by them as a basis for his lake-type system. Investigation of the bottom zone of lakes was for a long time one of the main objects of limnology.

The consumption of oxygen at the bottom during periods of stagnation is so slight in oligotrophic lakes that oxygen is never a minimal factor for the animals on the bottom. In eutrophic lakes, on the other hand, the oxygen in deep water may, some few weeks after the full circulation, diminish to a critical degree and ultimately may become an absolute factor. The result of this is that only those species of animals which can manage with quite small quantities of oxygen can populate the deep regions of eutrophic lakes. They live in these regions often in enormous numbers, whilst, in oligotrophic lakes, the fauna of the bottom is richer in species but poorer in numbers. The bottom zone of a lake can thus provide indications of its metabolic physiological character; in the bottom fauna of eutrophic lakes the genus *Chironomus* (= *Tendipes*) predominates, and in that of oligotrophic lakes the genus *Tanytarsus* and for this reason these lakes are called *Chironomus* lakes or *Tanytarsus* lakes respectively. Obviously this typification applies only to a small group of especially characteristic lakes, and all stages between the two extremes are found.

As well as the chironomid larvae other characteristic inhabitants of the profundal region are the larvae of *Sialis*, the alder-fly, and those of the midge *Chaoborus*, and also some molluscs, especially *Pisidium*, the pea-mussel, and oligochaetes and nematodes, some water-mites, ostracods, Harpactidae, and the developmental stages of some Cyclopidae. As can be seen from this list, all the organisms on the bottom do not pass their whole lives in the profundal zone. The insect larvae become winged aerial insects. But even the larvae of the chironomid *Sergentia coracina* in Lake Titisee in the Black Forest rise up into the pelagial region when, during the stagnation period, the oxygen content at the bottom falls below 2 mg/l (Wülker, 1961) and the larvae of *Chaoborus* make vertical migrations; they live during the day on the uppermost layers of the mud and at night rise to the surface of the water where they live a purely planktonic life (Berg, 1937; Schröder, 1961; Northcote, 1964); the light conditions, however, determine these migrations. The developmental stages of the Cyclopidae mentioned above pass a resting stage in the mud at the bottom which is more or less prolonged and is probably induced by photoperiodicity (Einsle, 1964) and then may be found in enormous numbers up to some millions per square metre; in the winter and spring they develop further.

Aids for the study of the fauna of the bottom may be divided into three groups—bottom grabs, core-samplers and mud-borers, and also dredges. It is difficult to take representative samples of the bottom because the fauna of the bottom is not evenly distributed and because especially the chironomid larvae lie half-passively on the bottom and are distributed over it by water currents (Barthelmess, 1963).

1. Bottom Grabs

These were developed for the purpose of bringing to the surface quantitative portions of the mud at the bottom as undisturbed as possible. By undisturbed we mean that the natural layering of the sediment and the bottom fauna should remain as far as possible intact. Briefly the grabs work on the following principle: most of these excavators have two opposed movable shovels. They are let down open on a line or a wire cable to the bottom, into which they sink by their own considerable weight. Then they are closed, i.e. the shovels operate with the greatest vigour against each other and thus excavate a definite quantity of sediment out of the bottom. The grabs are closed either by a drop-weight or by dropping the grab on the bottom and then pulling on the line. The closed grab is pulled up to the surface and the sediment is emptied out of it.

This method of using grabs involves the following sources of error:

1. The grab takes out only a square area of the uppermost layer of sediment and not, on the other hand, one of the deeper part out of which the shovels take a semicircular bite; the sample of sediment is therefore not qualitatively equivalent in all the layers of it. This fault has been overcome by the box-grab of Reineck (see below), but it applies to all other grabs in use. An additional fault of a similar kind is the fact that a grab that is manageable to a certain degree digs out only a relatively small part of the bottom surface; with the Ekman–Birge grab, this is 225–250 cm^2. Because, as has been mentioned, the bottom fauna is not in any way evenly distributed, there is the risk that erroneous conclusions about the population of the profundal region may be drawn from samples that are too few. As a basis for calculations at least ten grab samples at each depth are necessary from which an average is calculated. Riedl (1956) accurately watched the putting of the Petersen grab

on the bottom and saw that it partly swept the layers aside just before it reached the uppermost layers of mud, i.e. just the layers in which the micro-fauna lives. When a grab penetrates further in, coarser material is stirred up and this rushes out of the valves above. Riedl reckoned on a loss of 50–80% of the micro-fauna (ciliates, Turbellaria, mites, Copepoda, ostracods). This defect is therefore a serious one because it cannot be estimated with certainty. The error can only be kept as small as possible by releasing the grab slowly.

2. A further defect results from the variable consistency and composition of the sediment and from the closing mechanism of the grabs. The grab may sink very deep into very soft sediment and may even pierce through the uppermost layer. Grabs which close themselves when they reach the bottom operate in this respect very unreliably because the least resistance by the bottom operates the closure mechanism. Thus grabs of this kind are hardly suitable for soft bottoms. However, with grabs that are closed by a drop-weight, the decision to let down the drop-weight depends on the slack on the draw line, i.e. on the position of the grab on the bottom. All these facts must be taken into consideration.

There are difficulties on hard bottoms also. The lighter grabs do not settle correctly but lie more or less flat, do not sink in deeply enough, and no longer work quantitatively. Grabs which close automatically are here better because they have a more favourable grab action. On coarsely granular and stony sediment, stones may get jammed between the excavating shovels so that these do not close firmly, and part of the sediment is lost when the grab is pulled up. The loss of material increases as the depth through which the grab is drawn up increases, and it is difficult to guard against this. Such samples must naturally be rejected and there is nothing else to do but to work on until one succeeds in obtaining a sufficient number of serviceable catches. The development of bottom-grab technique and of the improvement of grabs themselves depend especially on knowledge

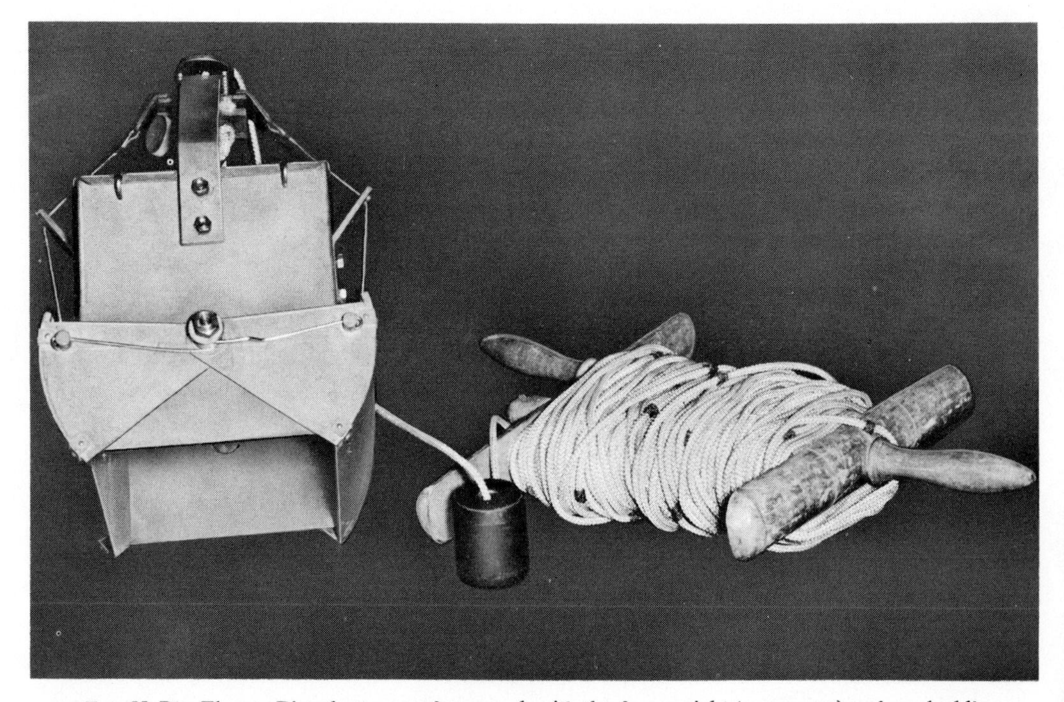

FIG. 55. The Ekman–Birge bottom grab, opened, with the drop-weight (messenger) and marked line.

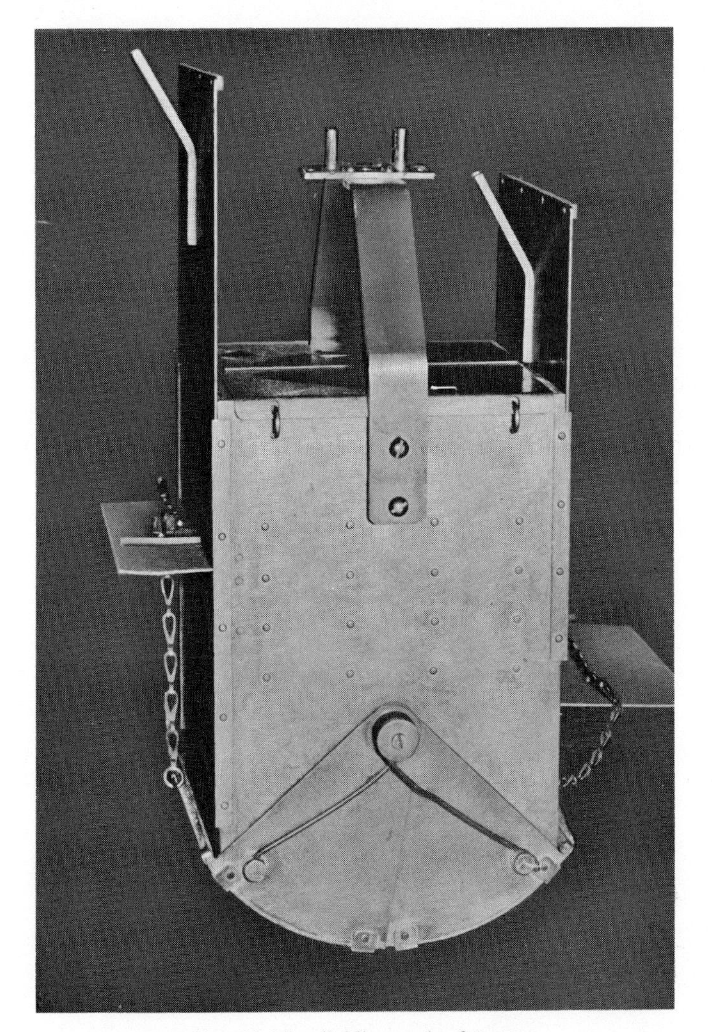

FIG. 56. The dividing grab of Lenz.

of the defects only briefly discussed here. Alsterberg (1922), Alm (1922), Wasmund (1932), and recently Riedl (1956) and Sander (1957) have been especially concerned with these problems.

Description of some types of grabs

The first bottom grab designed for quantitative work was devised by Ekman (1911) and improved by Birge (1922). It is still in general use in hydrobiology today (Fig. 55). Its two shovels, kept open against very strong spring traction by means of two chains, are closed from above by means of a drop-weight. One can readily detect when this has happened by a jerk on the line holding the grab. When it has happened one pulls the grab out of the bottom by careful traction on the line and pulls it up. Here again it is wise to use marked lines in order to be able to read off the depth of the catch at the time. The grab should be pretty heavy and it is best made of brass (not iron), because this does not rust in the water.

Its upper part is box-shaped and is closed by two movable covers which fall in under the pressure of the water when the grab is let down; they protect the sediment when the grab is pulled up before this is emptied out. Lundbeck (1936) increased the weight of the grab still more by means of superimposed layers lead added to it and in this way made it possible for it to penetrate also into firm sediment. The grab has a basal surface of 225–250 cm², so that the conversion to 1 m² is very simple by multiplication by 40. Wasmund (1932) made the grab bigger and heavier and adapted it to Ekman's original design.

Lenz (1931) introduced a very useful improvement of the Ekman–Birge grab. His dividing grab (Fig. 56) is rather taller, but has the same basal surface. Slots are made in two opposite sides at the same height and metal plates can be put into these. By means of them the contents of the box can be divided into horizontal layers and these can be separately studied. With this device, which is used everywhere, Lenz has obtained important results in the study of the vertical distribution of the bottom fauna in the deep sediment of lakes. A similar horizontal division of mud samples can be made with the mud-sampler of Jenkin and Elgmork (see p. 92).

FIG. 57. Auerbach's bottom grab: 1, open; 2, closed. f_2, c_1, c_2 = the closing mechanisms and the rod carriers; g = the ring-weight; a = the upper lateral walls of the box, loaded with lead; b_1 = the shovels, which in 2 have closed the box. (After Auerbach, 1953.)

For quantitative work in the deep marginal Alpine lakes with mainly clay sediment, Auerbach (1953) devised a box grab which works very reliably (Fig. 57). As its basal surface is only 100 cm², it is much smaller than the Ekman–Birge grab but, as it weighs 13·25 kg, it is much heavier. In contrast to the Ekman–Birge grab it is not closed by a drop-weight, but the shovels (Fig. 57, b_1) close automatically when the grab reaches the bottom. This is achieved first by the downward movement of an iron ring, which presses the shovels into the bottom (Fig. 57*g*) and, secondly, by a pull on the line when the grab is pulled up which completely closes the shovels. This grab takes completely smooth and undisturbed samples of the sediment from the bottom and also works very accurately in depths of more than 100 m.

Franklin and Anderson (1961) made a grab that works in a similar way but has no significant advantages over the Auerbach grab. Wasmund (1932) recommends especially

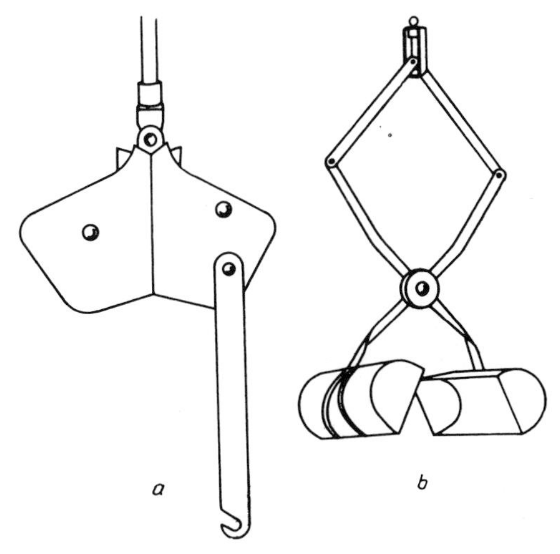

FIG. 58. Leger's Monaco grab (a); Ruttner's scissor grab (b). (From Wasmund, 1932.)

for sandy–stony sediment in, for example, high mountain lakes, the Monaco grab (Fig. 58) devised by Leger, the shovels of which automatically close when the grab reaches the bottom. The very reliable grab of Van Veen (1936) works on the same principle; it is held open by a small bar. When the grab reaches the bottom the two shovels spread out so that the bar is released. The draw line is attached in such a way that the stronger the pull from above is, the tighter the shovels close. A small model (1/20 m²) of the grab originally made for making marine biological investigations works very well for the study of the bottom in hydrobiological work (Fig. 59).

The box grab II of Reineck (1963) is positively worthy of mention. The penetrating box let down into the bottom has a basal surface of 22 by 30 cm, a height of 45 cm, and is interchangeable. It can be made heavier by added weights according to the solidity of the bottom. When it is hauled up the box is closed by horizontal claws so that a completely cubic sample of the sediment is removed. Because the grab is very large and very heavy (750 kg) its use is hardly possible in freshwater biology. The smaller apparatus, the 150 kg box grab, can only be sent down from the larger boats. Still smaller grabs can be made on the same principle.

Because this grab is very heavy it is difficult to take it on long excursions. Günther (1963) described a folding bottom grab measuring 25 by 17 by 6 cm which weighs only 1·5 kg and which can easily be taken on excursions. It works on the shovel principle, but its shovels consist of metal frames with a bag of coarse linen for collecting the sediment

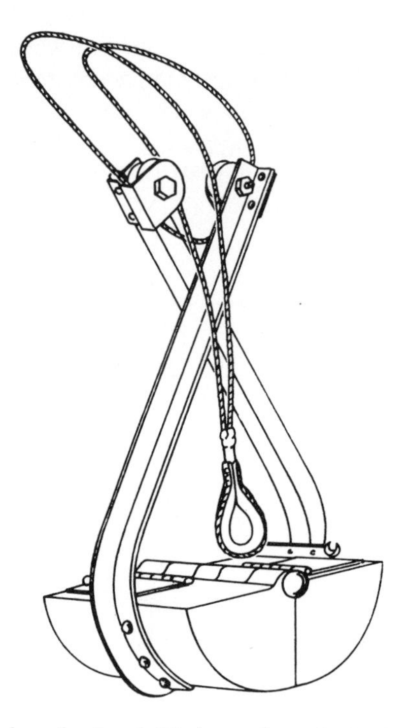

FIG. 59. Van Veen's grab, closed. (Redrawn from the Hydrobios catalogue.)

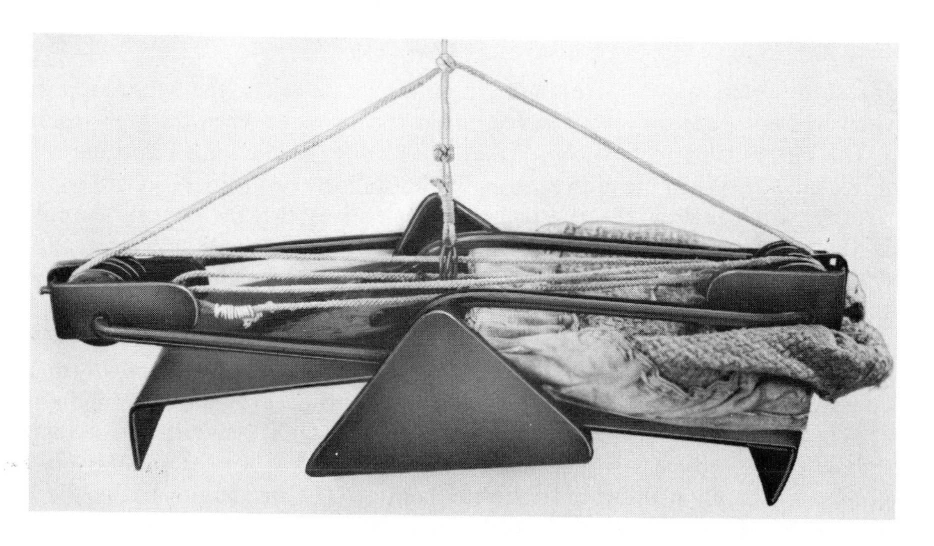

FIG. 60. Günther's folding bottom grab, opened.

and a cover fitted to them. The draw line passes over two rollers on each side which operate like the fixed and movable rollers of a pulley system (Figs. 60 and 61). The grab is weighted on each side by triangular lead weights which, when the grab is open, are suspended on a draw line. This suspensory device is disengaged when the grab reaches the sediment.

FIG. 61. Günther's grab, closed.

A pull from above operates the lateral draw lines of the pulley system and this closes the grab. The force with which the pincers of the grab close is always twice as great as the weight of the object seized. The makers state that there are only two possibilities—"the breaking of the cord on which the grab hangs or the recovery of the object collected". This, however, also holds for the Monaco and Van Veen grabs.

We should mention here, in addition the grabs of Lang (1930), Knudsen (1927), and Hentschel's improved Petersen grab, the grabs of Friedinger and the Ruttner grab as well the very interesting grab recently devised by Rieth (1960), which offers a number of advantages.

2. Mud-borers, Core-samplers, and Mud-samplers

These devices have been developed not so much for the study of the bottom fauna, as were the bottom grabs, but for taking samples of the deeper layers of the sediment that are as accurate as possible, i.e. undisturbed samples. Naumann has forcibly pointed out that the sediment, especially the sediment of the contact zone between the sediment and the water, is of the greatest significance for the metabolic processes in a lake.

All mud-samplers and core-samplers used today derive from the models of Apstein and Naumann, but Apstein's mud-sampler is hardly used nowadays. It consists of a brass tube 50 cm long with a diameter of 3 cm which is loaded at the top with a heavy lead weight. A valve at its upper end allows the water in the tube to escape upwards when the instrument is let down to the bottom and it closes again when the tube is pulled up. The lead weight drives the tube firmly into the mud, so that a profile of the sediment is cut out. The sample is later removed from the tube with a rod, but the layering of the sediment is easily disturbed by doing this. The Naumann sampler more often used nowadays delivers small but undisturbed cores. The tube made of hard glass consists of half tubes which can be rotated on each other and can easily be opened; this makes it possible to remove sediment more carefully.

For accurate investigation of the micro-layers in the mud and their bacterial flora, Liebmann (1950) devised a core-sampler the brass tube of which has, at intervals of 1 mm, bore-holes with a diameter of 0·5 mm. Before the sample is taken these holes are sealed up with adhesive tape (Sellotape, Scotch tape). After the sample has been taken in the usual way, the Tesafilm seal is perforated with fine pipettes and samples are taken for investigation at any required level of the sediment profile. After the sample has been taken the tubes are closed with a flap valve.

Züllig (1953) developed a similar instrument. The tubes made of plexiglass (perspex, lucite) have in their upper part holes at intervals of 20 cm for taking water out of the contact layer and in their lower part holes which permit the taking of sediment at intervals of 1 cm.

In this connection we should also mention the so-called bottom water-bottles which collect samples of water from the contact zone between the water and the bottom at greater depths. Such buckets have been used by Elster (1957), Liepolt (1960), and Murray (1962). Züllig (1959) modified for the same purpose the Friedinger sampler (see Fig. 30, p. 46). The cylinder is only 20 cm tall and it is closed by a drop-weight. For samples to be taken immediately above the bottom, the bottle is screwed on to a frame provided with a ring base and this rests on the sediment and ensures that the sampler lies only 20–30 cm above the surface of the mud. Elgmork (1962) has developed a very interesting mud-sampler from the Friedinger bottle, with which he has studied the distribution of cyclopid copepodites and the larvae of *Chaoborus* in the sediment of Swedish lakes. A metal disc is inserted below the two flap covers and into a central depression in this a metal rod can be inserted. A hole is made in the lower flap cover the diameter of which is the same as that of the metal rod. The open sampler is sunk into the soft sediment and is closed by the drop-weight. After the full cylinder is drawn up, the metal rod is led through the opening in the lower cover against the movable metal disc, and this is pushed upwards. In this way the sediment is pushed out of the sampler upwards, so that whatever portions of the sediment are required can be separated off with a metal plate. The cylinder must therefore be made perfectly smooth. According to Elgmork this sampler has yielded good results at lake depths up to 150 m.

Still to be mentioned are other models of core-samplers, among others those made by Perfilew (1927, 1929) and Reissinger (1930). Mortimer (1942) described under the name "Jenkin surface mud-sampler" a complex but very useful mud sampler. Frey (1954) and Deevey (1954) used, for the study of historical deposits in lakes, piston borers, which are used in researches on marine deposits, and produced core-samples several metres deep. They extend back as far as geological times and provide information about the history of the lakes (see also Frey, 1963). Züllig (1956) has taken very deep core profiles from Lake Zürich with Kullenberg's piston borer.

3. Dredges

The most important types of dredges have already been described in the section on the littoral region (pp. 77 ff.). Naturally they can also be let down into the profundal region to fetch up bottom material. Here three other models, which are especially useful at the greater depths, will be described: the "bottom-skinner" of Frolander and Pratt (1962), the sledge dredge of Elster (1933), and the closed dredge of Riedl (1955, 1960).

The bottom-skinner does not collect any sediment but brings up, by means of a horizontal net, the ground water immediately above the mud layer. The framework to which the net is attached is like a sleigh, which is pulled along the sediment. The net is the same as the net used by Clarke and Bumpus in their plankton sampler (see p. 45). The opening of the net lies only a few centimetres above the bottom (Fig. 62). So that a constant pull from above does not lift the dredge from the bottom, heavy globular lead weights are attached to the sides of the framework in front. Although the method has all the faults of a net catch,

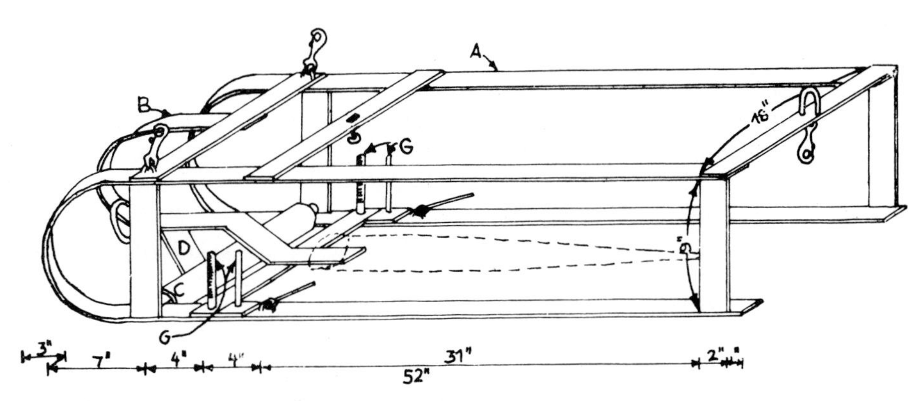

Fig. 62. Frolander and Pratt's bottom-skinner. The weights omitted; the net shown by dotted lines. (From Frolander and Pratt 1962, somewhat modified.) *A*, *B*, *D*, rods; *C*, the roller which facilitates the movement of the sledge; *G*, bolts for attachment of weights or lines. Particulars of sizes in inches.

the authors have obtained good results up to depths of about 12 m. But the net becomes choked up relatively quickly so that it is difficult to make exact statements about the amount of water filtered. This apparatus will collect the animals living in the uppermost layer of mud and also in the open water, which are not so well obtained with either bottom grabs or plankton nets or water-bottles.

The sledge dredge of Elster (1933) was made in order to collect sufficient numbers of the eggs of the blue char from the greatest depths of the Bodensee as this was not possible with ordinary dredges. The dredge consists of two metal plates, made like skis, 75 cm long and

15 cm broad, and turned up somewhat in front, so that the dredge may slide smoothly directly over the surface of the lake bottom (Fig. 63). To the middle of each ski a weight of about 7 kg is attached to prevent the dredge from rising up as it slides along. For the same purpose a foil which regulates the height is mounted above the frame of the net; this is a simple rectangular metal plate measuring 22 by 32 cm. A few centimetres in front of the opening into the net an iron rake is fixed across from ski to ski; its teeth slope forwards at

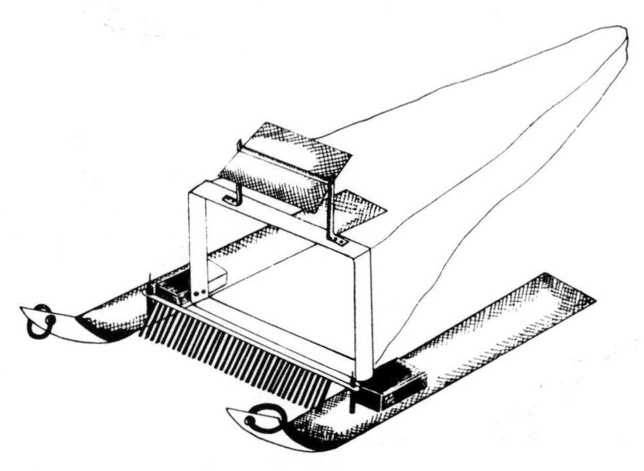

FIG. 63. Elster's sledge dredge. (Redrawn from Elster, 1933.)

intervals of 1 cm and end at about 2 cm below the bed of the sledge. This rake has two advantages—it stirs up the uppermost layer of the mud immediately in front of the opening into the net so that organisms are driven upward and fish eggs are whirled up and are more certain to get into the net. Further, the rake pushes larger obstructions aside and this protects the net. The opening into the net is quadrangular and measures 30 by 50 cm. The bag of the net is 1·4 m long and is made of wire gauze with a mesh width of 2 mm and is rigidly attached to the net frame by a flat sheet of iron which runs along the lower surface of the net. The detachable bucket which is attached to the net by a bayonet clasp has a diameter of 7 cm.

The sledge dredge, like the model of Frolander and Pratt, can only be towed along by a motor-boat and can be pulled up only with an efficient winch. A tow lasting 10 min (at a speed of 1·8 km/hr) yielded up to 175 blue char eggs from a depth of 160–200 m in the Bodensee, and innumerable specimens of *Pisidium*, up to 65 *Gammarus*, and numerous *Niphargus*. In contrast to this, ten samples taken with the Ekman–Birge grab collected only one fish egg (as well as other organisms). The dredge, however, does not operate strictly quantitatively because its collecting efficiency changes with increasing choking of the net. If one raises, with appropriate supports, the frame of the net a little, one can also collect the animals swimming immediately above the sediment, e.g. water-mites or the recently hatched larvae of blue char. The sledge dredge thus operates rather like the bottom-skinner of Frolander and Pratt. Riedl's (1955) sledge dredge collects quantitatively. It consists of a sledge dredge, a bag in the middle with detachable nets and a calculating device, which indicates the distance travelled by the dredge. This dredge can be closed after it has travelled a certain distance by a closing mechanism.

The closing dredge travels on broad skids. It has one anterior and two main shovels which are movable and collect the bottom material up to a depth of 5 cm. Since there is a space of 50 cm between the main shovel and the dredge, the coarse material in front of its mouth sinks back to the bottom and the dredge takes up only the lighter detritus and most of the microfauna. The main shovels can be retracted so far that they are 10 cm above the surface of the sediment; the dredge thus collects only from the layers of water near the bottom. Between this dredge and the bag of the net an intermediate bag made of coarse sailcloth (canvas) is inserted, to which the bag of the net is attached. This simple attachment has the advantage that the net can be quickly changed.

The most interesting feature of this dredge is the distance calculator and the closing mechanism. The distance calculator is connected with a spring steel spatula to a paddle wheel that passes over the bottom. The shovels in the apparatus described by Riedl penetrate to a depth of 5 cm into the bottom, but they can jump over stones and run over hard surfaces. The seat of trouble in this dredge (see Riedl, 1955) lies in the unreliable action of the paddle wheel. In the movement of the paddle wheel a cord from an inbuilt spool is unwound off and on to the middle narrower part of the axle and thus marks off the distance travelled. Closure is effected by two plates held open by a peg in front of the net opening which are disengaged and closed by means of a stop (knot, peg, etc.) on the cord. Because the gear reduction of the distance travelled by the dredge to the unwound cord is as 1:40, the stop on the cord is placed at 1 m distance in front of the stopping peg for the closing valves if it is intended that the dredge should travel 40 m. For details and a plan of the construction, see Riedl (1955, 1960). A rather simplified apparatus is used in the institute for lake research in Langenargen (communication by Dr. J.-U. Rixen).

4. Further Treatment of the Bottom Samples

The further treatment of the bottom samples taken with the bottom grab or dredge starts with sieving or washing of the sediment, this being usually done on the spot. To do it, nets or wire sieves are used, and one can easily make these oneself out of a circular or quadrangular wooden frame, over one side of which wire gauze is stretched. It is best to put a coarse sieve with a mesh size of 0·5–0·6 mm and a fine one with a mesh width of less than 0·2 mm over one another (see below). Kolkwitz's mud sieve is very useful. It is a quadrangular wooden box into the bottom of which a sieve made of brass of various mesh widths is inserted. Also useful are the metal sieves used for the analysis of sediment which are placed above one another in such a way that the coarsest sieve is on the top and the finer ones below it (see Fig. 13). However, the larger sieve nests are not convenient to handle on board. Rawson (1953) and Fremling (1961) use for sieving water a rotated bucket with a sieve at its greatest diameter as well as two lateral window sieves and a propeller which increased the efficiency of the apparatus still more.

Jonasson (1955, 1958) carried out thorough investigations into the mesh width necessary to filter out completely all animals except the Protozoa when quantitative bottom samples are being sieved. He found that when the mesh width was reduced from 0·62 to 0·51 mm, the quota of *Tanytarsus* larvae (Chironomidae) in the sieve residue rose to 47%. Further research showed that with a mesh width of 0·6 mm only specimens of *Pisidium* and the cocoons of *Tubifex*, the larvae of *Chaoborus* and the mature larvae and pupae of *Chironomus anthracinus* were retained quantitatively. Tubificidae and immature larvae of *Chironomus* were retained only by sieves with a mesh width of 0·2 mm. Sander (1957) came to the same conclusions. In contrast to sieves with a mesh width of 0·51 mm, sieves with a mesh width of 0·26 mm caught 111% more organisms and those with mesh width of 0·2 mm 246% more. Therefore in quantitative studies of the bottom, especially in problems of population dynamics in which immature larvae are of importance, a mesh width of less than 0·2 mm must be used, and in other respects the mesh width must be carefully adapted to size of the animals selected.

Jonasson (1958) himself, depending on Kolkwitz, used a quadrilateral sieve box (Fig. 64) which was 30 by 40 by 17 cm. Over the bottom phosphorbronze gauze was stretched which can be exchanged at will for other sieves with a narrower mesh. For this purpose the bottom surface is fixed to the box with angle irons, and a rubber seal makes it absolutely water-tight. The sieve is manipulated so that the bottom of the box is repeatedly immersed in the water; in this way the fine mud particles are washed out and only the sieve residue with the organisms remain behind in the sieve.

In general it can be said about sieves that the sieves made of fabrics work better than those made of metal. It is therefore more advisable to use a scraper on a handle covered with narrow meshed gauze for sieving the bottom sample.

Naturally the bottom samples can be sieved much more accurately in the laboratory. This is done, as has already been described, by several sieves placed above one another as in the method for treating mud. Löffler (1961) briefly described an automatic method for mud treatment (Figs. 65 and 66). His very simple apparatus is made of plexiglass and consists essentially of a detachable cylinder made in separate sections with a filter attachment which rests on a square base. A nozzle in the lower part of the cylinder will receive a water-inlet pipe and also a siphon which sucks off the column of water to the corresponding level of water in the cylinder. Several filters made of bronze gauze are fitted into the cylinder in such a way that the coarsest ones are above and the finest below. The cylinder can be quickly broken up into its sieve sectors which are screwed into one another and sealed with rubber rings. The sample is emptied into the uppermost sieve, and water is let in from below. The water rises up and is drawn off again by the siphon. The uppermost filter is flooded by the constantly rising and falling water column, and the finer particles in the sieve sink down to the lower sieves. Above the uppermost bronze filter, exactly at the height of the maximum water-level, there is a filter made of bolting cloth. The function of this is to catch and immerse particles adhering to the surface. The rate at which the water rises can be regulated by the water-tap and that of its fall by the length and internal diameter of the siphon. It is best to make the siphon tube so that it can be lengthened or shortened without loss of time. The water flowing out can be conducted through a narrow-meshed gauze filter which holds back the smallest forms. It is especially favourable if the filter is fixed not flat, but slanting and, in addition, folded in the cylinder (Fig. 66). The water-outlet discharges into the neck of a drop bottle, which, when it is closed, fits smoothly into the inner wall of the cylinder. With this arrangement it is quite easy to take the cylinder to pieces.

Only preserved samples should be treated in this manner. Fixation hardens the organisms so that they cannot be so easily damaged by the sorting out. In itself the method works very effectively.

Sorting the organisms from the sieve residue

The sieving procedure yields a sieve residue composed of sediment and organisms of appropriate size. The next step is to collect the organisms. It is simplest to work with a pipette and forceps. Portions of the sieve residue are put into a white bowl or a wide Petri dish over a white background and are covered with water. If the organisms are still alive, the collection of them is simpler than it is in preserved samples, which are very unpleasant because of the formalin. Part of the sample must be examined under the binocular microscope for the smallest organisms; this must be especially done with the fine sieved material from narrow-meshed sieves.

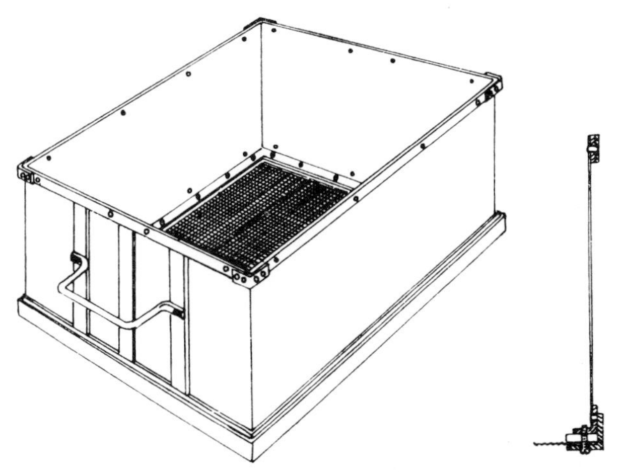

FIG. 64. Box for sieving sediment samples. (After Jonasson.)

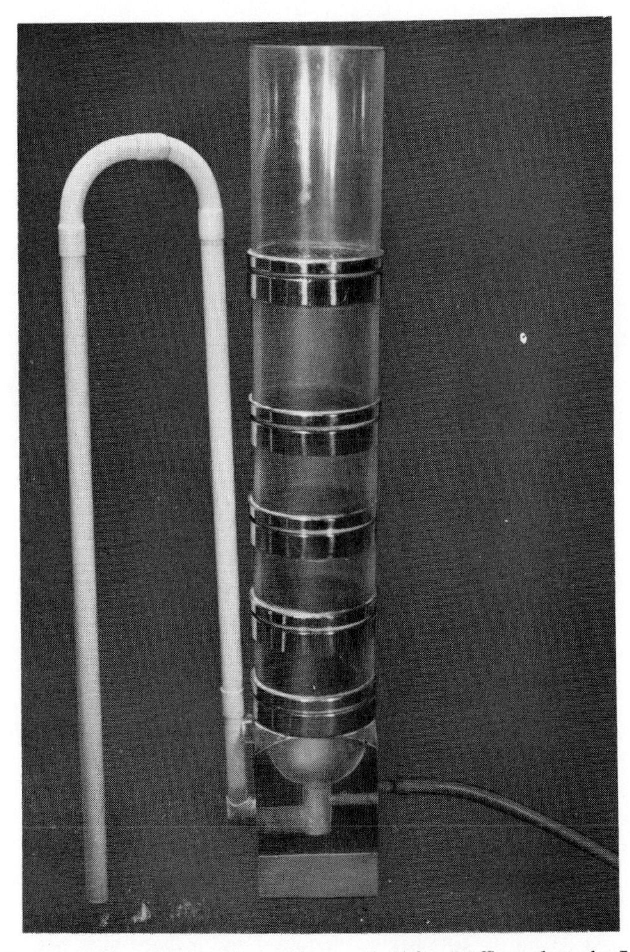

FIG. 65. Automatic apparatus for sieving mud. (After Löffler; photo by Löffler.)

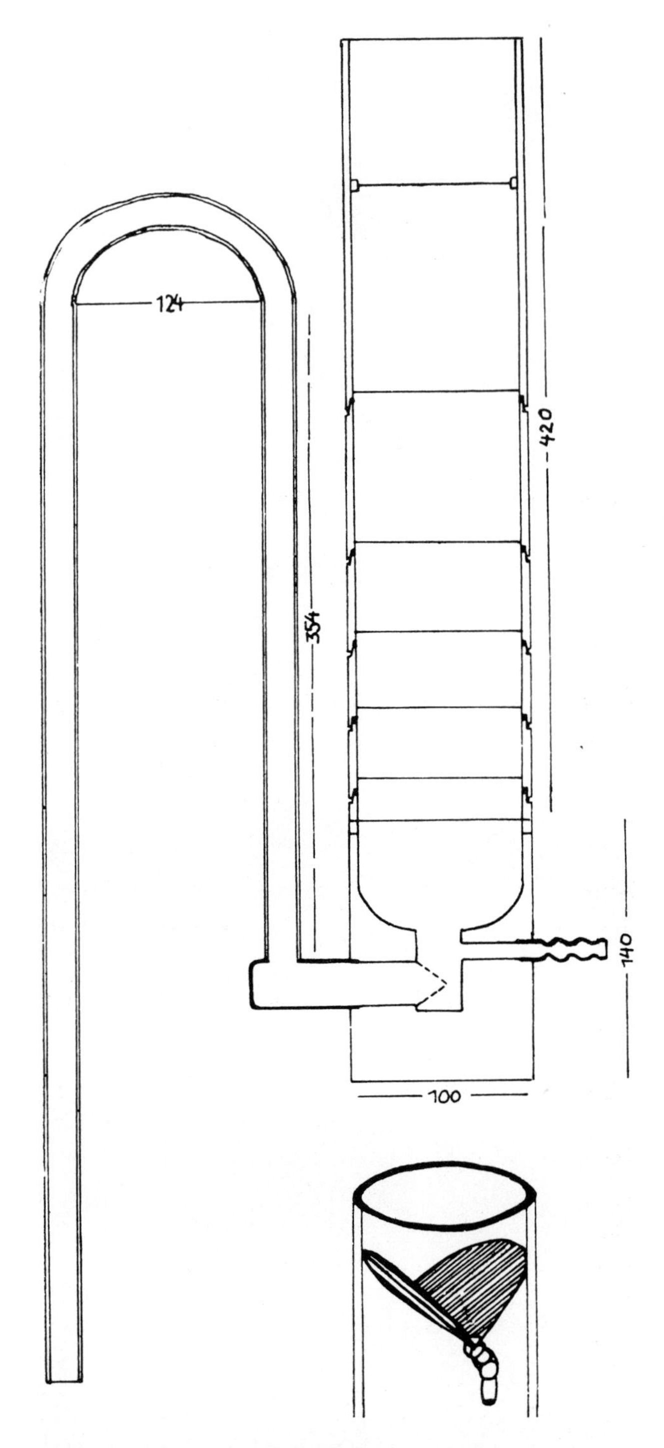

FIG. 66. Section through Löffler's mud-sieving apparatus. The water is led in on the right and is siphoned off on the left; below, the folded and inclined filter (see text).

This process of collecting by hand is very time-consuming. So far as I can see, two other methods are available for direct collection of the organisms or at least for concentrating them, i.e. working with specific denser solutions and working with an interrupted direct electric current. Neither of these methods requires elaborate apparatus.

Specific denser media

The principle of this method is the following. The specific gravity of aquatic organisms is rather greater than that of the water so that the organisms sink to the bottom if they do not swim—which bottom-living organisms rarely do. If one increases the specific gravity of the water by adding a specific denser solution, the organisms become relatively lighter and come up to the surface, as wood does in water.

The specific denser media used are solutions of high molecular weight substances, but not solutions of inorganic salts such as, for example, calcium chloride ($CaCl_2$) (Macan, 1958), which are, ironically, too active and penetrate too rapidly into the animals. After treatment with calcium chloride with a specific gravity of 1·1, the animals fall to the bottom after 15–20 sec, and in doing this lose much substance (Hell, 1960). When water-glass with a specific gravity of 1·15 was used, the animals swam all day at the surface (Hell, 1960). Anderson (1959) obtained similar favourable results with a solution of sucrose with a specific weight of 1·12.

The method of working is as follows. The sediment sample fixed with 40% formalin (with 40% formalin there is, in contrast to weaker solutions of formalin, no loss of weight) is put into a small bowl with the corresponding solution and the whole is well stirred; the animals quickly come to the surface. The supernatant solution, together with the animals, is poured through a funnel the inside of which is coated with a piece of coarse net gauze. The fluid is collected and used again. The gauze with the animals collected in it is washed with water until the last remains of water-glass or sugar solutions are washed out. The Trichoptera with stony cases and the molluscs with shells, which are naturally both too heavy to reach the surface, are picked individually out of the sediment with forceps and transferred to the fixation fluid. A sample of sediment is checked in 20–25 min. To make the sugar solution, about 300 g of sugar are dissolved in 1 l. of water.

Working with an interrupted direct current

This method gives satisfactory results when only small portions of mud are selected. It is best to use a quadrangular glass trough 80–100 mm long, 30–40 mm broad, and about the same height. Small metal plates as electrodes are attached to the ends of the glass trough. As a source of current, Dittmar (1951) used 120 V direct current (Fig. 67). If the current flows uniformly many species respond too slowly to it. To obtain a better effect, the current is interrupted at certain intervals and then a special stimulus is applied each time the circuit is interrupted. The power supply is therefore simply interrupted by a Morse key with 10–20 Hertz (cycles per second) (Fig. 67).

The layer of mud put into the trough should be not deeper than 1–2 cm, and the water above it should be 1 cm deep. The animals living in the mud flee from their environment when the current is switched on and come to the surface and place themselves transversely to the direction of the current. The larger animals (Trichoptera, Ephemeroptera, Plecoptera, Coleoptera, and larvae of Nematocera) appear there, according to Dittman's experiments, after a few seconds; Oligochaetes appear in about 15 sec and Nematodes and Protozoa first after 30–40 sec. It follows that with the same current density the effect (current flow) on the smaller animals is weaker. In addition the covering of the body plays a part: soft-skinned animals respond more strongly than those with body armour or those carrying cases.

The larger animals that reach the surface are picked out with forceps and the smaller ones with a pipette. Then from time to time the whole is stirred up and allowed to settle and the remaining animals are collected under the microscope.

Because this is a reaction of living animals, naturally only samples of living animals can be taken in this manner; and animals that die during the procedure are not collected.

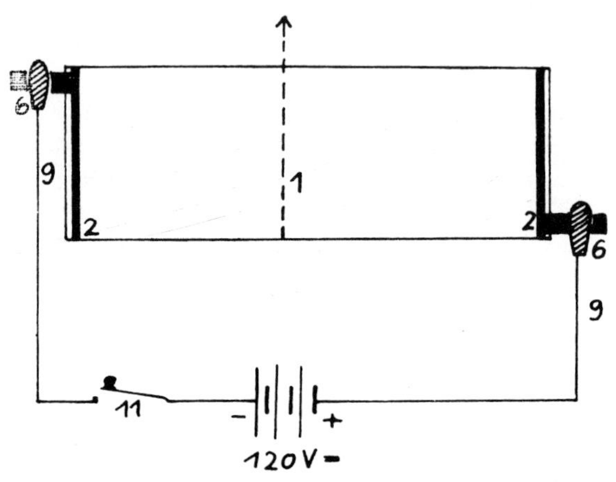

FIG. 67. Apparatus for the use of an interrupted direct current for the collection of organisms from samples of mud. 1, cuvette; 2, electrodes; 9, their leads; 6, connection terminals; 11, the Morse key. (After Dittmar.)

Statement of the results

The population determined by studies of the bottom fauna is commonly expressed either as the number of organisms of each species or groups or as the fresh weight of all individuals of a species or group, relative to a unit of surface.

The number of organisms per unit of surface can be shown graphically or in the form of a table. A graphical statement is significant only if the different colonization of individual lakes is clearly to be compared, as, for example, Fig. 68 shows for some lakes in the upper Black Forest. The groups of organisms in the reference areas, which are always of the same size, are recorded as schematic contour figures (sketches) of characteristic representatives corresponding to their number. The pictures thus constructed can be directly compared quantitatively with one another. Nevertheless, a summary of the results in the form of tables is essential for more accurate statements.

For the estimation of the productivity of the profundal region it is important to give, in addition to the numbers, a calculation of the fresh weight of all living animals in each surface unit.

The enumeration of individuals does not provide comparable values for calculations of production because the biomass of individual organisms varies a great deal. Production must therefore always be related to volume or weight. The wet weight, dry weight, and the ash weight are important. The fresh weight is determined by weighing the animals which no longer give off moisture on blotting paper. To estimate the dry weight the animals are dried at 80° for 4 days and then weighed, and then ashed at 600° for 1 hr and 20 min (Slack,

1955). Subtraction of the ash weight from the total dry weight gives the dry weight of organic material. Neil (1937–8) obtained the dry weight of animals that he had dried for 3 hr at 120°, but there are no investigations into what are the best conditions for the treatment of animals for the estimation of their dry weight.

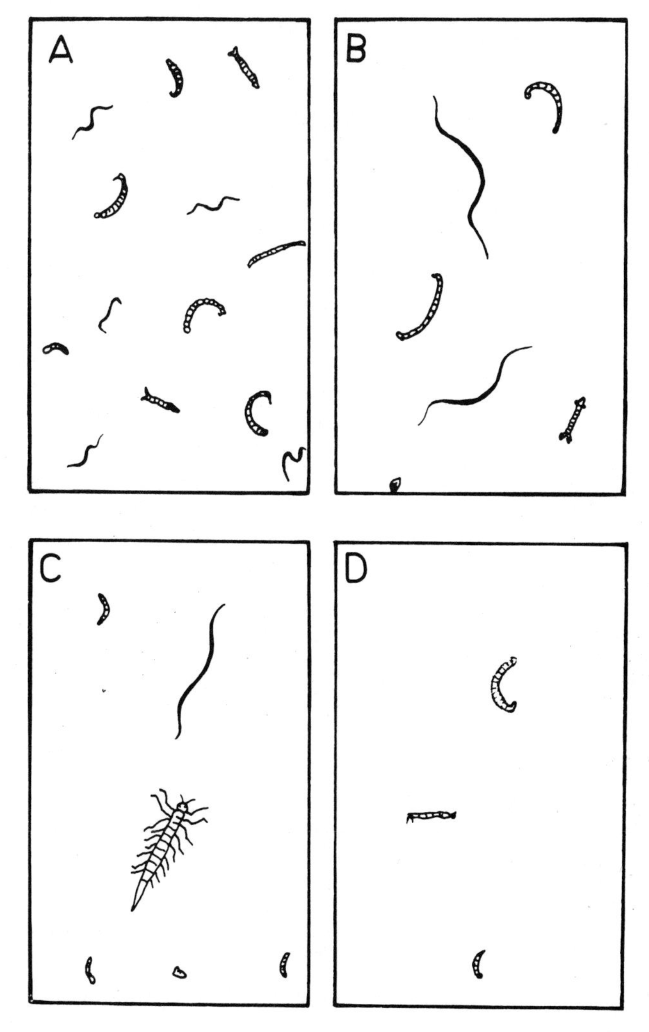

FIG. 68. Population of the bottom of various lakes in the upper Black Forest (after Lundbeck). A, Feldsee 10 m; B, Titisee 25 m; C, Titisee 4 m; D, Windfällpool 3 m. (see text).

Lellak (1961, 1961a) calculated that the annual production of chironomids in some Bohemian pools amounted to 240–1050 kg/hectare and that it supplied one-third to a half of the food requirements of the fish.

To estimate the fresh weight the organisms must first be preserved in 40% formalin solution (not in 4%) or in alcohol, because with these there is greater loss of substance (Shadin). Even when 40% solution of formalin is used, the estimation of the weight, in comparison with that of the living animals turns out to be rather too low (Hell, 1960).

V. THE QUANTITATIVE COLLECTION OF THE IMAGINES EMERGING IN THE BENTHAL REGION

It must be again noted here that the chironomid larvae constitute an important part of the population of the benthos of standing water. The full-grown larvae pupate and the winged midges emerge from the pupae. This often happens in a very limited period of time and under definite weather conditions. In the Titisee the sergentias *(Sergentia coracina)*, for example, emerge only during $1\frac{1}{2}$ months after the break up of the ice in April, and no imagines are found during the rest of the year (Wülker, 1961). The adult midges of *Tendipes anthracinus* emerge in the Danish Lake Esrom within 3 days and only at night between 7 p.m. and 1 a.m. (Jonasson, 1961).

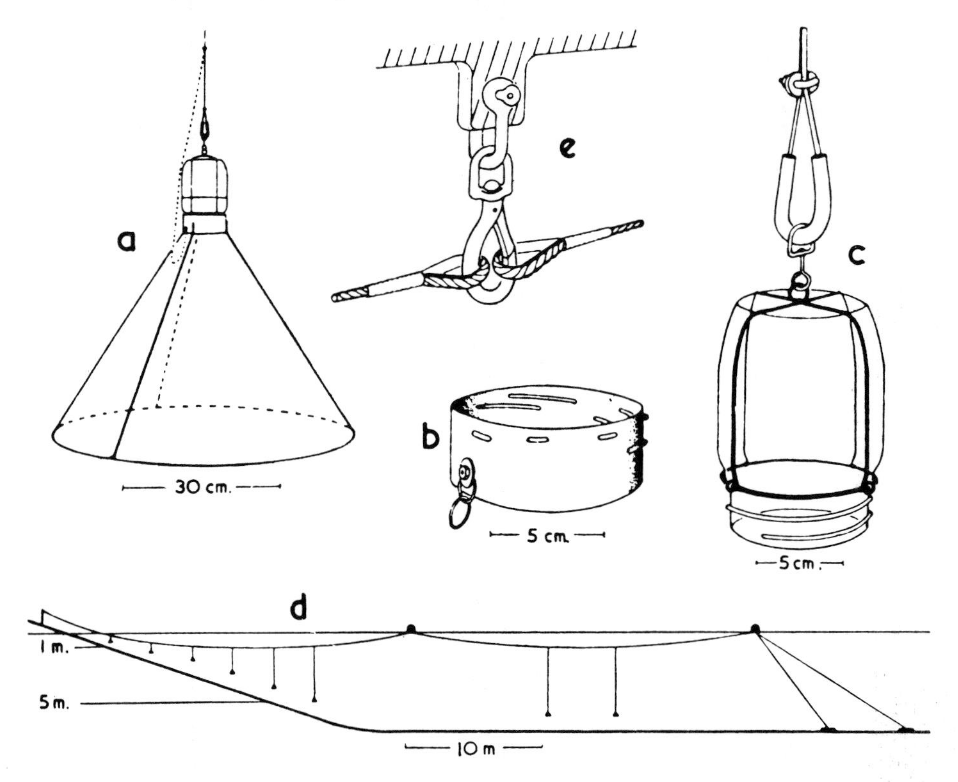

FIG. 69. Construction and use of funnel traps. (From Mundie.)

The number of animals emerging over a bottom area in a certain period of time is determined by funnel traps. These funnels are conical or pyramidal and are exposed under the water close to the bottom. Funnels which project beyond the water surface are exposed to the wind and waves and do not give accurate results. Brundin (1949) first used underwater traps of this kind, and later Palmen (1953) and Jonasson (1954) also used them. The idea was derived from Russian workers.

It is best to make the frame of the funnel of brass (Fig. 69) and to cover it all round with bronze gauze. Its surface opening has an area of 0·25 m^2 and its ring has a radius of 28·2 cm. Its height is 42 cm and its lateral supports are about 50·6 cm long. The mesh width selected for the gauze depends on the smallest organisms to be caught. Mundie used 0·25 mm. The funnel is closed above by a strip of copper sheet (Fig. 69b) to which the bronze gauze is soldered.

As a vessel for catching the emerging imagines an ordinary honey or marmalade jar is used, attached to the funnel with its opening downwards (Fig. 69c). To do this holes are bored in the copper sheet and pieces 5–6 cm long of 2 mm strong copper wire are threaded through these holes (as shown in Fig. 69b). The jar can be screwed into this wire frame and must be tested for a time until it fits. Figures 69a and c show how the funnel is suspended.

The funnels can be exposed singly or in series, as Fig. 69d shows. They must be examined at least twice a week so that the pupae and imagines are not damaged by attacks of fungus. There is still cont roversy about how far above the bottom the funnels should be exposed; for this and about other details, see Mundie (1956).

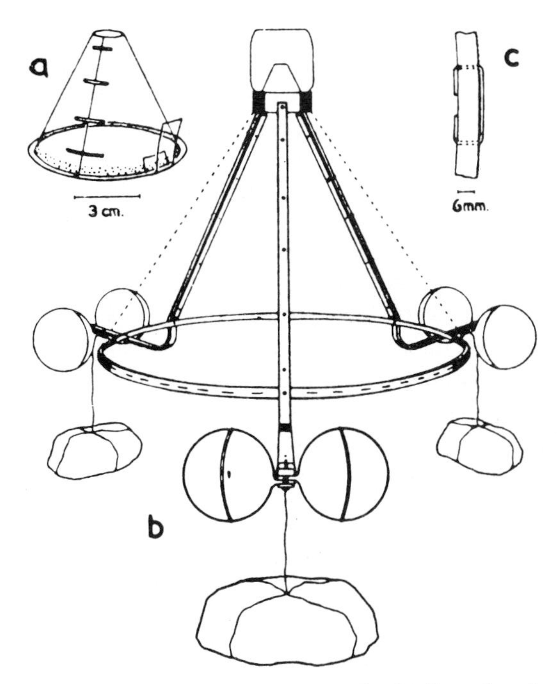

FIG. 70. Construction of a funnel trap and insertion in the littoral region. (From Mundie.)

A similar funnel, covered with nylon gauze, is used in the shallow water of the littoral region. A smaller funnel, made of celluloid, projecting into the collecting bottle prevents the insects from being washed out of the bottle again by the force of the waves (Fig. 70a). The funnel is exposed partly immersed. Its stability is ensured, on the one hand, by buoys, and, on the other, by stones with which it is anchored to the bottom (Fig. 70b).

Recently Schlee (1965) constructed a "trap roundabout" (revolving collector) which is put on to a funnel trap and consists of two rotating circular discs. The discs rotate so rapidly that one collecting vessel stays for as long as an hour above the funnel opening before it is replaced by another. By this method the daily rhythm of emergence can also be determined.

For funnel traps used in running water see p. 122.

CHAPTER 4

METHODS FOR THE INVESTIGATION OF RUNNING WATER

IT HAS already been emphasized in the introduction that running waters are, in contrast to lakes, open systems, the characteristic features of which are their flow and constant transport of materials in one direction. Odum (1957) and Teal (1957) were the first to analyse accurately the balance of matter and energy in such a system.

The movement of the water makes demands on the inhabitants of the running water and these have partly led to very specialized adaptations. Free-swimming organisms can maintain themselves for long periods only in slowly running waters, e.g. in the lower courses of larger streams. Among the free-swimming animals only some fishes can maintain themselves in fast-flowing waters against the thrust of the water by muscular power, their bodies being hydrodynamically especially favourable to doing this. The number of swimmers increases as the rate of flow of the stream decreases. Most of the organisms in rapidly flowing streams belong to the bottom fauna and maintain themselves against the movement of the water by anchoring themselves to a substrate. The higher plants root themselves on the bottom, mosses attach themselves to stones, and algae form crusts—difficult to detach—and overgrowths on stones and on the higher plants.

Some animals live on smooth stones on which the action of suckers or sucker-like devices (*Liponeura*, *Epeorus*, Fig. 71) may be of great importance; but only these live actually in the strongest current. Others seek out places sheltered from the current; they live in the mats of mosses and other densely growing plants or between the stones on the bottom in the so-called "dead water" in which the reduced movement of the water is biologically very significant. Finally, many organisms escape the strongest currents by taking refuge in the larger and smaller spaces in the gravel of the stream, or close to the bed of the stream in the so-called "hyporheic sphere of life" (Orghidan). The methodological difficulties of the study of running waters result from these various kinds of behaviour of the organisms and from the necessity of bearing in mind all the structural elements in the bottom.

The current, in proportion to the generally decreasing force of the water downstream, sorts out the sediment according to particle sizes and the particle size usually decreases downstream. The animal and plant communities also change downstream in correspondence with the variable behaviour of the organism in the stream in relation to the substrate.

These biological variations in running water are influenced, as well as by the current and structure of the bed of the stream, by the temperature of the water, and this depends especially on the extreme values. The annual range of temperature increases downstream from the source, and this is also basically true of the daily range of temperature, but it depends also on local conditions such as shading or the entry of cooler water. Many organisms do not withstand the high summer temperatures of the lower reaches of the stream, while others are more tolerant of the temperature. These facts make it possible to understand that the temperature conditions in streams are a prerequisite for the understanding of their populations. Therefore the determination of the temperature amplitude during a long period of time is especially important.

Thienemann (1912) and Steinmann (1907) were the first to recognize accurately the influence of the movement of the water, the bed of the stream, and the temperature on the composition of the living communities in a river. Already, more than 60 years ago, the fishery biologists divided the stream from its source downwards into three regions according

FIG. 71. The larva of the ephemerid fly *Epeorus*. (Magnified about 1·2 times.)

to the fish characteristic of them. The uppermost region is the salmonid region with trout and grayling; then follows the barbel region with the barbel; and, finally, the bream region with the bream as the characteristic fish. The fact that these fish live in such different regions of the stream depends on their different temperature tolerances and especially on their reproductive habits. During the last 10 years especially, biological investigations of the fish have established a general zonation of the living communities in streams with the result that nowadays the upper mountainous region of a river is called the rhitron (instead of the salmonid region) and the lower reaches in the plains, the potamon (Illies, 1961).

The methodology in general: if one follows the changes in the temperature and the gradient and structure of the bed of the stream downstream from its sources, one finds that the conditions of life change very rapidly but also remain the same over more extensive regions of the stream. In conformity with this the change of fauna also occurs over increasingly greater stretches of the stream. It is, however, important for accurate investigation of the stream that the sites of the investigation must be situated closer together in the upper reaches and especially near the source than those selected further downstream. Because an accurate knowledge of the living conditions in brooks and streams is important, every biological investigation should include determination of the temperature, analysis of the substratum according to the kind of stones and particle size, estimation of the current speed, and determination of the chemical data. These methods of investigation are described on pp. 10 ff. For a biological analysis one must make a clear decision whether a qualitative investigation is sufficient or whether à quantitative one is necessary, and this depends on the problem in hand at the time. Quantitative methodology is still very unsatisfactory. In every instance the variable relation of the organisms to the substratum must be borne in mind.

I. INVESTIGATION OF THE PLANT POPULATION

It is a question of algae: diatoms, blue–green and green algae, also bacteria and fungi, mosses and the higher aquatic plants, which may themselves be substrates for the lower plants.

The qualitative collection of the phanerogams and mosses is very simple; in small streams they can usually be taken directly from stones or from the bottom. In deeper waters they must be taken with dredges or plant hooks (cf. Standing Water, pp. 77 ff.).

To obtain the algae which live on the stalks and leaves and in the cushions of plants, it is sufficient for qualitative work to pull out the plant cushions and then to wash them out.

Bacteria *(Sphaerotilus)* and fungi *(Fusarium* and some others) occur only in polluted waters and can be easily scraped off with a knife from stones and the higher plants. It is more difficult to collect the algae from stones. Filamentous species, such as, for example, *Ulothrix*, *Vaucheria*, *Stigeoclonium*, etc., can be easily scraped off stones; on the other hand, many diatoms and blue–green and green algae and also the bright red *Hildenbrandia* form quite thin growths on stones over which the current flows (epilithic). These species can only be collected by vigorously brushing the stones or scraping them with a knife. For accurate investigations the various surface of the stones must be treated separately, attention being paid to their position in relation to the current. The samples obtained are put into glass tubes or wide-necked flasks and are best fixed at once with 4% solution of formalin or Pfeiffer's mixture (see Appendix I).

Quantitative collection

Quantitative methods are difficult because the samples must now be related to a unit of volume or surface area. The growth area of mosses and the higher aquatic plants can be directly measured, but in the estimation of the biomass it must be borne in mind that the submerged plants such as *Fontinalis*, *Callitriche*, *Ranunculus*, *Glyceria*, etc., have a small growth area but in proportion to it a large mass of vegetation. The statements are made in grams of fresh or dry weight related to a unit of area, e.g. 1 m^2 (see p. 83).

Sphaerotilus and also the fungus *Fusarium* can be scraped off a surface of stone selected at random, e.g. one of 10 cm^2. The amounts thus obtained are also stated in grams of fresh and dry weight, and it must be remembered that the estimation of the fresh weight is inaccurate because of the water adherent to them. Filamentous algae can be treated in the same way as bacteria, fungi, and mosses.

Method of Douglas

Barbara Douglas (1958) has described the following method (Fig. 72) for the biocoenoses of epilithic algae. She takes a steel rod hollowed out at one end to a depth of about 5 mm and cements into this hollow some nylon bristles. The cement is composed of 2 parts of resin dissolved in 1 part of hot castor oil. Wire bristles are not suitable because they scrape off too much from the stone. The nylon bristles should be 0·5 mm thick and 3·6 mm long.

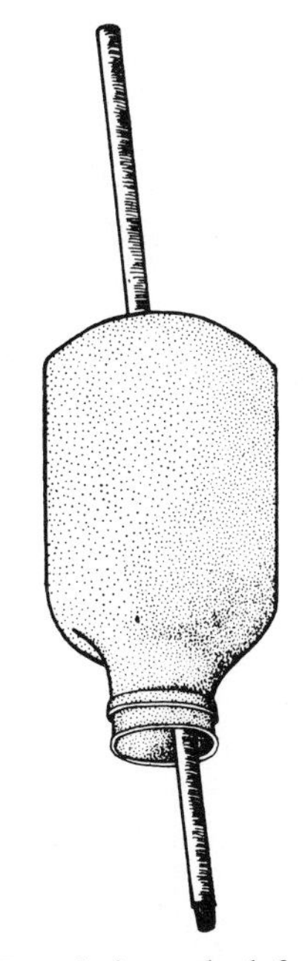

FIG. 72. Douglas's method for collecting epilithic algae. (Redrawn from Douglas, 1958.)

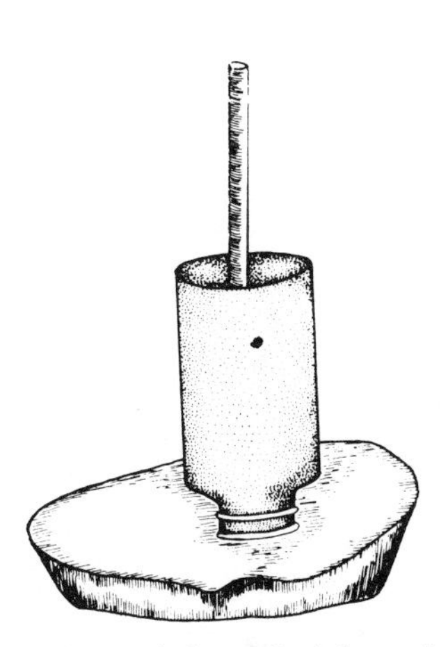

FIG. 73. Douglas's method for collecting epilithic algae. (Redrawn from Douglas, 1958.)

Longer bristles bend up and shorter ones do not get into all the surface irregularities of the stone. The surface to be quantitatively brushed is marked out through the neck of an inverted 50 ml polyethylene flask the bottom of which has been cut out. As Fig. 73 shows, the area of the opening of the neck of the flask (about 2·5 cm^2) is scraped clean from the stone with the bristles. The algae are taken up with a fine pipette and put into a collecting flask, and finally, the polyethylene flask and also the bristles are carefully washed and the water used for their washing is also put into the collecting flask. Samples of this kind are taken from several similar substrates and the areas sampled are added up so that a representative sample is obtained relative to the whole area investigated. The simple apparatus is easily made according to this description by Douglas.

Method of Margalef

Another very interesting method has been devised by Margalef (1949, 1949a). He removes small stones from the running water or knocks small pieces off rocks, fixes them in 4% formalin solution, and stains them with Delafield's haematoxylin. Then they are washed in distilled water and passed through a series of alcohols into a mixture of absolute alcohol and ether. This procedure requires about half a day. The stones are now removed from the solution, but must NOT be dried, but a solution of collodion in alcohol–ether is dropped on to them. This is allowed to dry and then pieces of the collodion film can be detached and put on a microscope slide. If the material is well dehydrated it can be immediately embedded in Canada balsam. This method has the advantage over the Douglas method that it preserves the natural position of the algae.

Here may also be mentioned the methods of Kann and Golubič, described on p. 73.

Douglas's method for algae on solid rock (underwater)

This is a modification of this authoress's method mentioned above, but the algae removed by the bristles must be protected from being washed away. This is achieved by means of a firm flap, which with a thick rubber ring sits on the rock so that it protects the surface scraped. The steel rod is hollow so that the algae detached can be continuously sucked up into the collecting flask (Fig. 74).

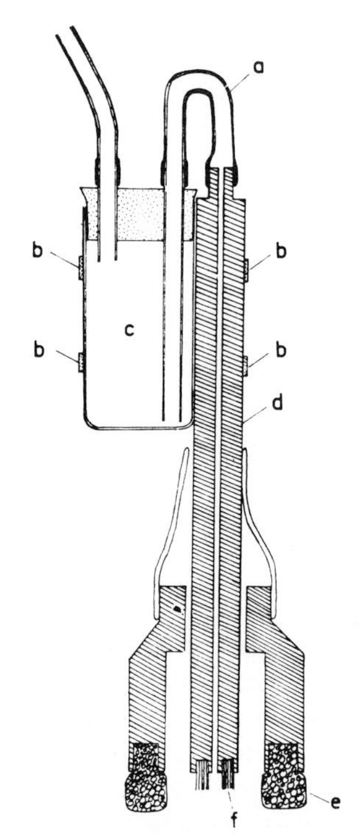

FIG. 74. Douglas's method for obtaining epilithic algae under water. The algae are scraped off with the nylon bristles *f*, attached to the hollow steel rod *d*, and are sucked through the glass tube *a* into the collecting vessel *c*, which is attached to the steel shaft by rubber bands *b*. The rubber flap *e* prevents the algae under the water being washed away. (Redrawn from Douglas, 1958.)

Douglas's method for collecting epiphytes on aquatic mosses (and flowering plants)

This also is difficult. The epiphytes can be obtained by squeezing out these cushions of moss and putting the water squeezed out into a collecting flask. But many organisms then remain on the substrate, and this must be remembered when strictly quantitative analyses are being made. Douglas has given the following methods.

The plants collected are flooded with water in a shallow Petri dish and the sand and other particles are removed; the water used for washing goes into the collecting flask. Finally, the volume of the moss is determined in a small measuring cylinder by estimating how much water it displaces. The water used for this also goes into the collecting flask. The mosses are then cut up into small pieces and put into a glass funnel with a long stem (Fig. 75). Inside

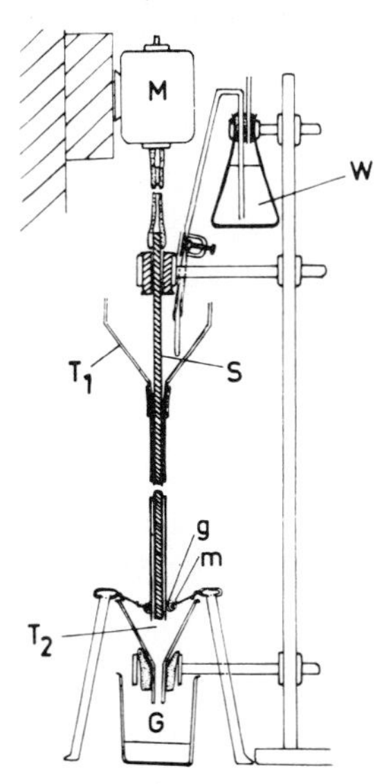

FIG. 75. Apparatus for breaking up mosses to obtain the algae growing on them. The electromotor M rotates the shaft S, which drives the mosses (not shown in the figure) in the funnel T_1, through a narrow tube into the funnel T_2, so that they are pulverized. From the vessel W water is dripped into T_1 so that the broken up mosses are washed into T_2 and the final collecting vessel G. g and m are rubber and wire fastenings. (Redrawn from Douglas, 1958.)

the stem of the funnel a glass rod rotates, between which and the inner surface of the stem of the funnel there is a space of 0·5 mm. In this way the cut-up moss is ground up and this powder is washed by drops of water added to the upper funnel into a second funnel below the other one which conducts it into the collecting flask. The glass rod can be rotated by an electric motor. The upper part of the glass rod, which passes directly to the first part of the stem of the funnel, must be grooved like a helical screw so that the mosses are driven downwards.

The counting of the cells can be done in the same way as for plankton (see p. 59).

II. INVESTIGATION OF THE ANIMAL POPULATION

Animals that live in running water behave very differently in relation to the substrate. Rarely do they live on the upper side of submerged stones (Figs. 71, 76, 77), but most of them seek out niches, clumps of moss which are sheltered from the current, the under sides of the stones, or the hyporheic system of spaces. Most of the difficulty in the investigation of running water is caused by having to catch the animals in all their "refuges" and in not being able to select some of these.

The smaller streams are always investigated upstream from their lower reaches. This ensures that organisms washed away when samples are taken are not recorded in another site of investigation to which they do not belong.

FIG. 76. The rheophilous mollusc *Ancylus fluviatilis* from above (lower series) and from the side (upper series). The current flows from the left.

FIG. 77. The pupae of the trichopteran *Agapetus*, firmly attached in their cases to the submerged stone.

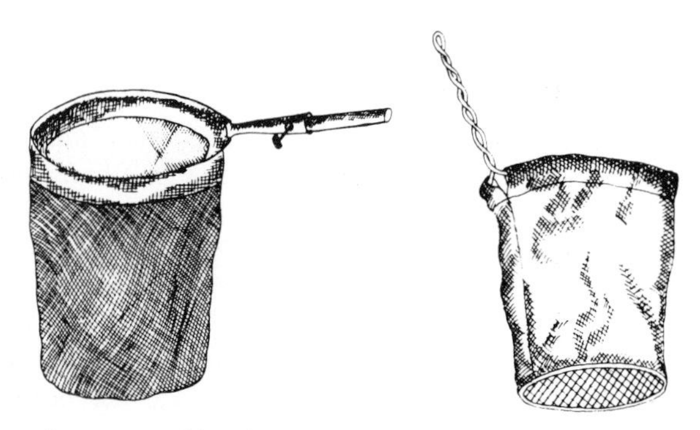

FIG. 78. Thienemann's compound hand-net; the sieve on the right is put into the net on the left.

1. The Population of Plant Mats

Plant mats offer shelter from the current and act filters in it. In them, therefore, live many organisms that do not expressly favour currents and which feed on the captured detritus. To obtain a picture of the content of organisms it is best to wash out thoroughly some plant mats in Thienemann's compound net. This consists of two nets, a larger hand-net made of nylon gauze and linen (Fig. 78), into which a smaller coarse metal sieve is placed. The plants are washed out in this sieve. The smaller organisms and the detritus thus pass into the gauze net, whilst the plants and the larger organisms remain behind in the sieve. These organisms can then be easily sorted out of the clumps of vegetation.

The plants freed of detritus and organisms are weighed but without the water adhering to them. To do this they are first dried on filter paper.

The organisms are sorted out of the detritus in a shallow dish, and the smallest ones are counted under a binocular. Because the detritus is the basic food of the whole living community, the amount of it must be ascertained as accurately as possible. Next the wet volume is determined (in a graduated cylinder without supernatant water). Then the detritus is dried for 24 hr at about 80°C and its weight is then determined. Finally, the dry substance is ashed and the residue is weighed; this consists of the mineral constitutents, which naturally constitute a large proportion because sand and fragments of stone are retained by the plants. The difference between the dry and the ash weights gives the amount of the organic constituents, i.e. those that the organisms can utilize.

The plant volume can also be given as a standard unit for the quantitative estimation. This volume can be determined by filling a graduated cylinder full of plants with as few interstices as possible, but clearly by doing this the spaces in which the organisms live are not taken into account. A "standard curve" of the relationship of the weight and volume should ideally be worked out for each species of plant under consideration in which one first determines the appropriate volume for each fresh weight and constructs a curve from the results, which also include the intermediate values, so that later on one has only to determine the weights.

It is impossible to estimate the actual surface area of substratum provided by the plants even if such an attempt were made. The relationship to the plant weight and volume, or, as a third possibility, to the colonization surface area in the water body, is only an approximate solution.

2. The Population of the Stones

The animals are either fixed to the stones, as are the larvae and pupae of the midges *Liponeura* and *Simulium* as well as the case-building larvae and pupae of some caddis flies (Fig. 77), or they hold on to the underside of the stone and quickly escape as soon as one lifts the stone from the water and turns it over. The animals try to reach the underside of the stone again and thus easily fall back again into the water. To this category belong especially some of the flattened larvae of some ephemerid flies of the genera *Epeorus*, *Ecdyonurus*, *Rhitrogena*, and *Heptagenia* (see Fig. 71).

The stones in small mountain streams and shallow rivers picked out directly with the hand must be immediately washed in a small hand net so that the mobile animals do not fall back into the water. Then those attached to the stones are carefully detached with forceps.

If the animals are not to be examined alive or kept alive for some time, then the whole sample should be immediately fixed and preserved. If, on the other hand, the

the animals are to be taken away alive, they are put into heat-insulated Dewar (thermos) flasks with only a very little water packed with plenty of moss and taken as quickly as possible to the laboratory where they are transferred to shallow, well-aerated, and cooled dishes.

In order to state the population of the stones quantitatively it is again related to the stone surface area as has already been described in the discussion of the surf region of the shore, and the method of Schräder (1932) is selected for doing this. The greatest breadth and length of the largest surface of the stone are multiplied together and are evaluated as the populated surface. The largest projection of the stone is thus its populated surface. In doing this it should be remembered that the position of the stone in relation to the current is decisive for its population. In general, however, it is true that the smallest projection of a stone is for hydrodynamic reasons directly opposed to the current and does not favour population; Schräder's method therefore provides approximately correct values. In each instance one should observe and note the positions of individual organisms on the stone and in relation to the current; from this, and not from statistical calculations, the most interesting problems arise. Thus the populations of the upper and lower sides of a stone are quite different, a fact which is not expressed by statistical calculations.

The population of the stone surface is finally calculated as population per square metre. An example will explain this. If the size of a stone is 20 by 15 by 7 cm, its greatest projection is $20 \times 15 = 300$ cm^2; if there are 257 organisms living on the stone the population density per square metre is 8567; statements of this kind can be compared with one another. For studies in production biology it is naturally more correct to state the weight of organisms per unit of population. In the calculation stated above one is naturally not content with the study of only one stone but considers, according to the size, 10–20 stones, so that the differences in the colonization are approximately compensated.

If one determines how many stones lie on a unit area of the bottom of the stream and determines, by the method described, the population of these, one can state the populations of the stones in relation to the bottom area (Alm, 1919).

Dittmar (1955), on the other hand, refers not directly to the bottom area, but considers the volume of the populated substrate and arrives at population units for the various kinds of substrate. To determine the unit of population of the stones, the volumes of all the stones lying on 1 m^2 is determined, a value which should be fairly constant. The organisms are collected from a part of the stones, and their number is then calculated to the total stone volume and is stated as the population per square metre. If, therefore, 1070 organisms are collected from a stone volume of 3·21, but if there is in 1 m^2 of bottom area a stone volume of 12·51, then one unit of stone population is 4180 organisms/m^2.

Here it may be remembered that Ehrenberg (1957) has also worked with stone volumes in the surf areas of lakes. Redeke (1923), Gejskes (1935), and Beling (1949) calculated, on the other hand, the actual surfaces of the stones with compasses and tape measures and stated for each species the population density per unit of surface.

It is difficult to take stones from larger flowing waters; if they are not very deep, it is possible with a pole-scraper. Otherwise there is only the dredge, with which the stones are collected with other sediments and can no longer be studied separately.

3. The Population of the Finer Sediment

It is difficult to collect the organisms which live on gravel, sand, and muddy bottoms quantitatively or qualitatively. Most of the methods, worked out for the smaller streams, operate in such a way that the bottom is stirred up and the specifically lighter organisms are washed by the current into a net held ready, before they sink to the bottom again. Other methods, the shovel-samplers, collect the sediment itself, which then must be elutriated. To investigate the sediment in larger streams a bottom grab or a dredge must be used, but these do not provide a satisfactory quantitative result. To investigate the hyporheic sphere of life, i.e. the community living between the particles, special methods are used.

Flotation procedures

For simple qualitative investigations one puts into the current as near as possible to the bottom, under the site of the investigation, a net made of coarse plankton gauze which has a quadrangular opening. In the choice of the mesh-width a compromise is necessary. If it is too small, the net soon becomes choked and the current drives the organisms out of it again; if it is larger, the smaller organisms such as water mites, ostracods, copepods, etc., pass through it and the catch is no longer either quantitatively or qualitatively representative. One abandons the smallest organisms and does not go below a mesh-width of 0·5 mm.

If the stones are to be separately investigated for their population, they are handled as described above. The bottom of the stream immediately in front of the net is first stirred up so that the organisms thus stirred up are swept by the current into the net. For the sources of error in the flotation procedures, see pp. 115 ff. Illies (1952), for example, has investigated the population of the River Mölle by this method. If the sediment is especially muddy, it can be freed of its finer particles in a narrowmeshed sieve (cf. Profundal, p. 96) and the organisms can be sorted out of the residue in the sieve.

The simple methods cited are naturally not very quantitative; for this, some safeguards are necessary which will be described below. One must understand in this connection that on the whole there are, so far, no fully satisfactory quantitative methods for the biological investigation of running water. About this Pleskot has written: "There are complaints everywhere about the inadequacy of the 'quantitative methods' for collection from stony bottoms, and no remedy for these has yet been found. It would not only be futile to hope that it could be possible to collect all the organisms in the collecting area, but there can also be no doubt that one does not always capture, either from catch to catch or from species, the same percentage of the animals actually present." This inadequacy of the methods is an unavoidable evil about which one must not delude oneself. Macan (1958) has summarized the methods, and Albrecht (1959) has critically compared them with one another.

Next some "quantitative" methods which operate according to flotation procedures will be described.

In shallow streams the sampling cylinder (Fig. 79) of Neil (1937) works very well. It consists of a cylinder open above and below, made of sheet iron (less effective, but still usable is a quadrangular box), the lower border of which is strengthened and sharpened so that the cylinder penetrates easily into the substrate. On the box there should be two handgrips by means of which one can more readily drive the apparatus by rotating movements into the bottom. The area investigated is bounded by the lower surface of the cylinder. A few centimetres above its bottom the cylinder has two openings measuring 15 by 15 cm, the anterior one through which the water flows being provided with a coarsemeshed grating, the posterior one having a fine net. Both these openings can be closed from above by a slide valve.

The apparatus works as follows. After it is placed in the bottom and with the slide valves closed, the stones are freed from the animals attached to them and the plants are washed; stones and plants are then removed from the cylinder. The substrate is then dug up to a depth of about 5 cm and it is raked through and vigorously stirred up. After about

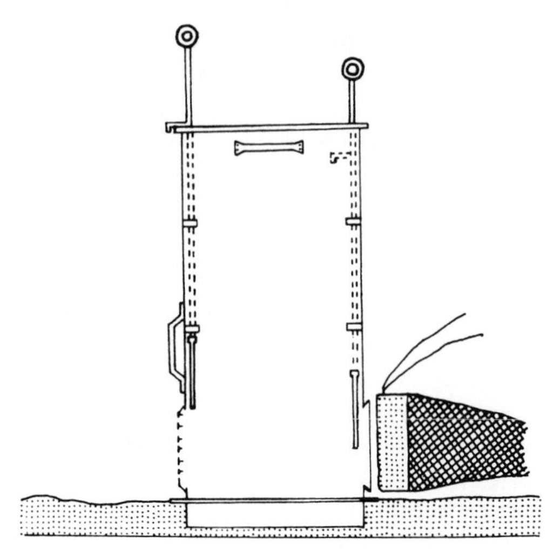

FIG. 79. Neil's trap-cylinder. (Redrawn from Macan.)

half a minute, after the coarsest particles have sunk to the bottom again, the slide valves are opened so that the water streaming through the cylinder drives the organisms into the posterior net. The raking of the substrate is repeated twice more and then the organisms should be practically quantitatively in the net.

By this method Needham (1927) first worked in a primitive manner and later Gejskes (1935) and Jonasson (1948). Neil improved the apparatus in the manner described and introduced the anterior and posterior slide valves. Wilding (1940) used a somewhat different container.

In shallower streams the square foot stream-bottom-sampler much used in the U.S.A. proves good. A vertical and a horizontal metal frame are firmly joined together (Fig. 80). A gauze net is fixed to the vertical sides and also a rod with which the sampler can be held in the stream. In operation the apparatus is held against the current and the bottom of the stream is stirred up. The organisms stirred up are swept by the current into the net.

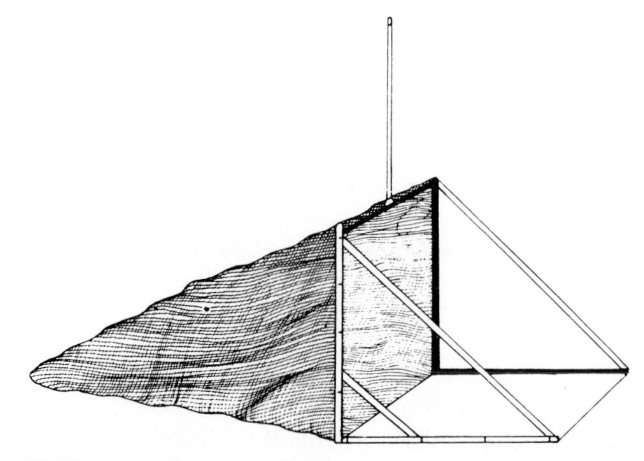

FIG. 80. The square foot stream-bottom-sampler. (Redrawn from Macan.)

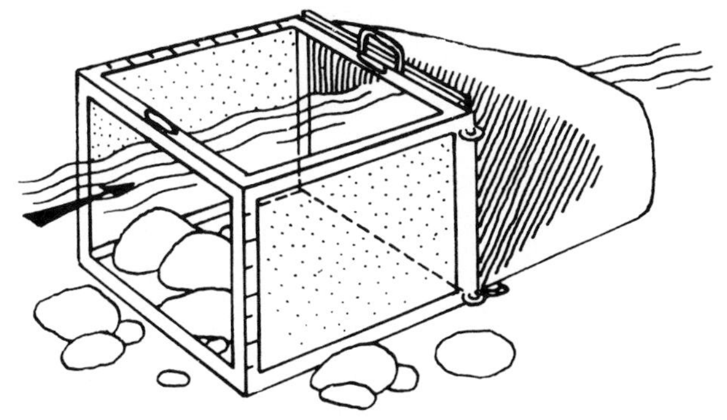

FIG. 81. Collecting apparatus of the Vienna-Kaisermühlen Federal Institute. (Redrawn from Slanina).

The Federal Institute for Water Biology and Effluent Investigation in Vienna works on the same principle but uses a quadrilateral box which is open above and in front, and canvas is stretched over the sides (Fig. 81). A net is fixed to the hinder frame and is protected by strong cloth. Between the box and the net there is a slide valve. It works as in the methods described.

Sources of error in flotation procedures

The procedures assume that all the organisms are set free from the substrate by stirring up the bottom of the stream and that all have the same chance of being washed into the net; but this is not so. Many of the smaller organisms especially hold fast to the heavier particles of the substrate and sink down with these within half a minute. The same is true of the large animals, which in general sink more quickly to the bottom. It is therefore, in contrast to the cylinder of Neil, better to allow the current to operate constantly and in this way to put up with a greater portion of the detritus. Many organisms in brooks, especially young inset larvae, live deeper than 5 cm in the sediment and are not captured by the flotation procedures if one does not stir up the substrate to a greater depth than this. For the choice of the correct net, see p. 95; in all instances the compromise mentioned above must be made.

Dredges and shovel methods (shovel-samplers)

These devices are moved against the current and take up in this way, to a depth of about 5 cm, both the substrate and also the organisms in a certain area. The area is calculated from the width of the anterior border of the shovel and the path covered by it.

The dredge of Usinger and Needham consists of an iron frame, measuring 4·5 by 12·5 by 7·5 cm (Fig. 82). At its anterior end there are bars fixed close to one another, their lower ends being pointed and bent a little forwards. When the dredge is drawn along, they dig into the bottom of the stream and stir it up, but hold back the larger stones. The dredge is pulled against the current on a line, so that the organisms stirred up are swept with the finer sediment into the net fixed to the posterior side of the frame. Further details can be seen in the figure.

The dredge has the advantage that it can also be put into deeper streams from a boat or a bridge. It is sunk down to the bottom of the stream and pulled upstream for a short distance. The area from which it collects is determined from the length of its anterior border, 45 cm, and the distance over which it is pulled.

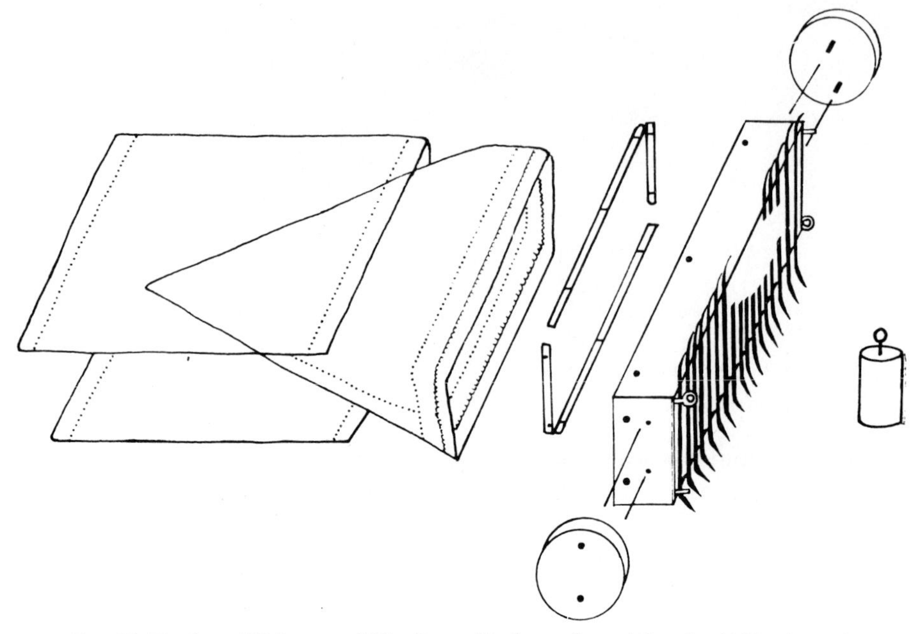

FIG. 82. Dredge of Usinger and Needham. (Redrawn from Albrecht, 1959, cf. text.)

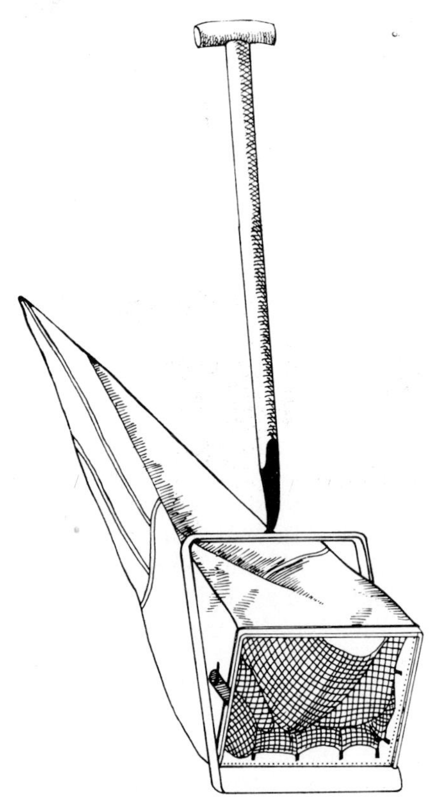

FIG. 83. The shovel-sampler of Allen and Macan. (Redrawn from Macan.)

The shovel-sampler of Allen and Macan is, on the other hand, pushed like a shovel forwards through the stream bottom, and it thus collects the sediment into two nets attached to the metal frame of the shovel (Fig. 83). A wide-meshed net holds back the coarse particles of the substrate and a fine-meshed net made of wire material picks up the fine sediment and the organisms. The anterior border of the metal frame is sharpened and the whole is further strengthened by lateral braces. An attachment made of metal gauze, the height of which is the same as the breadth of the shovel, prevents the organisms from being swept away over the shovel by the current. The apparatus is 22·5 cm broad and it takes up a layer of sediment 5 cm thick. The sampler can naturally be made broader, but then it is more difficult to handle. Allen (1940) made a similar sampler.

Dittmar (1955) made on a similar principle the stream-bottom-shovel and has investigated in detail with it the Aabach in Sauerland. According to his description (Fig. 84), this shovel consists of a shovel-like box 30 cm long and 30 cm broad over which fine gauze

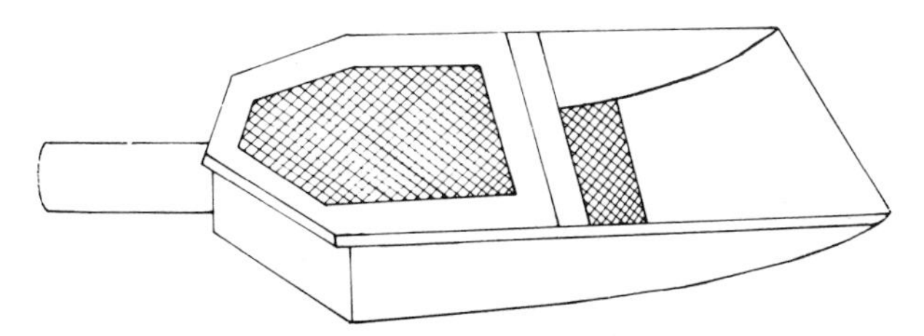

FIG. 84. Dittmar's stream-bottom-shovel

(bronze gauze with a mesh width of 0·1 mm is stretched, the anterior end of which is bevelled off and strengthened by a strong plate. Over half of the upper opening the same gauze is stretched. In the hinder wall there is an opening and a device for putting in the collecting vessel. This consists of an aluminium or tin box with a lid. From it the bottom is removed and its under side is made so that it can be screwed on to the closing device; in the lid a large hole is cut (almost to the edge of it) and then into this the gauze mentioned above is soldered. If one has several lids with different mesh widths, one can conveniently make a size selection of the organisms caught. In front of the opening in the hinder wall of the shovel a coarse sieve is placed in a suitable way by which stones, leaves, etc., cannot get into the collecting vessel.

The shovel is used as follows. The apparatus is grasped by both hands and its sharp anterior end is thrust into the substratum to be investigated and all stones, mosses, and the like taken along by it are pushed forwards until the hind end again lies over the starting point. With a quick jerk forwards and upwards, the apparatus is then lifted out of the water; in quiet water it is lightly shaken until all the fine particles have been washed away. Finally, stones are brushed off and removed from the apparatus. Mud particles should be finely powdered and washed; plants are carefully pulled to pieces and washed. All organisms are brought back to a more rapidly flowing point and into the collecting vessel by slapping on the water and by a rapid forward movement of the shovel. Finally, one can tilt up the apparatus and hold it up in order to flush out the last animals in the small can.

Percival and Whitehead (1929) had developed a similar apparatus.

Drawbacks and sources of error

The shovel-sampler described can hardly be used in stony streams because all the larger stones obstruct its movements; this is also true of Dittmar's stream-bottom-shovel as I have myself repeatedly proved. These forms of apparatus work, on the other hand, well on gravelly, sandy, or muddy bottoms. A further drawback is that they can only be used in shallow brooks. The use of nets involves the problem, already described, of the most favourable mesh width. Organisms that live deeper than 5 cm in the sediment—practically all the hyporheic fauna—are not captured.

The use of bottom grabs and dredges in bigger rivers

In order to take samples of the sediment from deeper waterbodies that are to some degree quantitative, we use bottom grabs and dredges. The dredge of Usinger and Needham has already been mentioned. Steinmann's simpler dredge (see Fig. 51, p. 78) is also good.

In larger rivers the Petersen grab can be used, as Behning has done in the River Volga. The grab weighs 40 kg and can only be used with a cable-winch from a large boat. It digs to a depth of 6–10 cm over a bottom area of 0·1 m^2.

Knöpp used the Ekman–Birge grab (see p. 86), but attached it—not to a cable—but to a brass rod which could be extended to a length of 5 m. In this way it was possible to insert the grab perpendicular into the bottom even with currents flowing at up to 1·5 m/sec and to a depth of 4·5 m. In shallower rivers it is possible to work with the grab attached to a rod in currents flowing even more rapidly, but the resistance of the bottom sets a limit to its use.

In the lower Elbe, Van Veen's grab (see Fig. 59, p. 90) has given good results (e.g. Kothé, 1961, and verbal communication from Professor Caspers, Hamburg); its use in brooks should also be tested.

4. The Hyporheic Fauna

The investigation of the hyporheic fauna, i.e. those organisms that live in the interstices between the particles of sediment under the bed of the stream and immediately on the shore of rivers in the zone where the ground water and the flowing water mix, is done by the method of Karaman and Chappuis by holes dug on the shore or in the islands in the river, or by driving in perforated metal tubes, or with the aid of exposed sand-tubes (see below).

Method of Karaman and Chappuis

A small hole is dug out in the sediment at drained places in the river (Fig. 85a) or on the shore of it (Fig. 85b). The sediment removed is repeatedly freed from mud in a bowl and the detritus is collected in a hand net made of gauze No. 16. Water collects in the hole and this percolates out of canals and bores in the sides of the hole. This water can be directly taken up with fine glass pipettes and investigated chemically. If the hole is full of water it is drained with a sampling vessel holding 0·5–1 l. in such a way that the detritus is also collected. To do this the sampling vessel is pushed, with its opening directed downwards, straight down to the bottom of the hole and is then slowly rotated; the suction draws the detritus into the vessel together with the water. The hole is drained to its last remaining content. The contents of the vessel are also put through the hand net made of gauze No. 16. It is best to put a coarse-wire sieve in front of the net so that the coarse gravel and stones do not get into the net and the organisms are not injured.

(a)

(b)

FIG. 85. (a) Hyporheic digging in an island of gravel in a brook; in the foreground the ladle and sieve. (b) Hyporheic digging on the shore of a brook. The stream flows on the upper edge of the picture.

The hole is drained 2–3 times in all. It is not in any way sufficient to remove the water in the hole with the hand net because then the organisms and the detritus at the bottom of the hole are not completely collected and the sample gives a bad yield.

Approximately quantitative catches, comparable with one another, are obtained if one digs out holes of the same size and ladles out the same quantities of water. Useful results have been obtained in researches in the Black Forest with holes measuring 25×40 cm (=0·1 m²) and with 15 l. of sample-water (Schwoerbel, 1961, 1961a).

Samples rich in detritus are put into shallow dishes with a little water. The larger organisms are picked out directly with a pipette, but controls must be done on the dishes for several days. The smaller organisms are counted under the binocular in a portion of the whole sample and the results are then converted to the whole of the sample (cf. p. 52).

Working with perforated tubes

Tubes made of iron or, better, brass or steel, are used, which are 50 cm long and have an internal diameter of about 5 cm. Their lower ends are sharpened and they also have holes 5 mm in diameter above the pointed part for a distance of 10 cm. They are driven into the bottom of the stream and after they have been exposed for 5–10 min their contents are sucked out of their upper ends with a water-pump. It is enough to take about 20 l. of water with each drilling. The water, together with the detritus, is filtered through the hand net and the sample is treated as has been described above.

The method has the advantage that the hyporheic fauna can also be investigated in parts of the stream in which the water flows over in which no holes can be dug. The samples, are, however, quantitatively not so rich as those taken from the holes, which is probably not due to the fact that the population would be smaller, but probably to the filtering effect of the sediment itself, which occurs when the water in the sediment is sucked out. By the use of the same tubes and considering the same amount of water, the catches are therefore comparable only to a limited degree between each other. In every case an analysis of the particle size is necessary; and this naturally also holds for the holes dug out.

Exposure of artificial sand tubes

The procedure described serves for general orientation with regard to the colonization of the hyporheic interstices and also for obtaining research material. Tubes full of sterile sand (Fig. 86) which are dug into chosen parts of the hyporheic sediment, are especially suitable for the investigation of the hyporheic biocoenoses. If these tubes are, in addition, filled with sands of different particle sizes and are exposed for different periods of time, they provide significant information on the importance of various environmental factors for the hyporheic biocoenosis.

The titrisol ampoules made by Messrs. Merck of Darmstadt are very suitable; they contain about 160 cm³. Both ends are cut off and the tubes are closed with the screw-tops of polyethylene wide-necked flasks. These covers are cut out and 1 mm nylon gauze is stretched over them; the best adhesive is UHU-plus. The interstitial fauna can pass through the 1 mm openings in the gauze as they please; naturally the same mesh width must be used for all sand fractions (the finer sand does not trickle out when it is moistened). If different particle sizes are exposed, it is best to mark the tubes with adhesive tape of various colours so that even in turbid water the correct particle size can be immediately recognized and can be collected (Fig. 86).

FIG. 86. Sand tubes which are dug into the hyporheic sediment (see the text).

The tubes collected from time to time are emptied and carefully freed of mud. The detritus is collected in nylon gauze No. 25 and put into a Petri dish, the bottom of which is marked with lines, and the organisms are quantitatively counted under a stereoscopic lens.

III. INVESTIGATION OF PLANKTON IN RUNNING WATER

The larger rivers and streams also contain plankton which partly comes from standing water in the headwaters of the river, but also partly develops in the river itself. In the middle Rhine the phytoplankton is so abundant that the oxygen content of the water is distinctly influenced by it (Knöpp, 1960).

To collect this plankton the apparatus usually used for the investigation of lakes (plankton nets, water-bottle, and pump methods) are used. The horizontal water-bottle, especially designed for collecting water samples from flowing water, has already been mentioned (see p. 46).

IV. DETERMINATION OF THE ORGANISMAL DRIFT

The organisms in running water, in so far as they do not belong to the plankton, are prevented from being carried away by the current by the structure of their bodies and their modes of life. However, it may naturally happen that an organism is taken up by the current and carried away by it. There is an increased risk of this "misfortune" when there is a marked increase in the movement of the water, e.g. a high water, and also if the animals leave their refuges which are protected from the current. Many organisms in rivers remain during the daytime under stones and in spaces in which the current is hardly strong enough to sweep them away. If they leave these refuges at night the risk of being washed away is increased. The actual diurnal and nocturnal variations in the number of drifting organisms has been determined and they can be explained in this way.

To determine the drift a plankton net, preferably one with a quadrangular opening, is put into the current and fixed there. It should neither touch the bottom nor project above the surface. The time during which it is exposed depends on the particular features of the current and the water-level; exposure for 0·5–3 hr is the right order, but then the nets must be changed. In this manner the drift over a period of 24 hr is investigated. This can be done at about monthly intervals. In this way one obtains a picture of the daily and seasonal variations in the drift. In choosing the net care should be taken not to use a net with a mesh width that is too small. The narrower the mesh width is, the shorter should be the duration of its exposure. The nets collected up are washed out, the organisms are identified and counted, and the amount of inorganic (mineral) constituents and also that of organic detritus is determined. The results are stated per hour of the exposure time or for greater intervals of time, continuously over 24 hr. The nets must be hauled in at the right times so that they do not become choked up and thus give values that are too low.

V. COLLECTION OF INSECTS FROM RUNNING WATER

The insects living as larvae and pupae in water mostly live as winged animals near the shores of brooks and rivers. They are especially the stone flies (Plecoptera), may flies (Ephemeroptera), caddis flies (Trichoptera), and some families of the two-winged flies (Diptera). They are stripped off the plants on the shore by a collecting net and collected from stones, the piles of bridges, etc. Many can also be collected from the stream and, in doing this, their special swarming periods should be borne in mind. Individual species emerge to fly about at different times. Tables about these can be found in the literature on the subject (see the identification literature). The flight period changes also in the same species according to the altitude and geographical latitude; this must be remembered when the fauna is being evaluated.

The animals collected are not dried but are preserved in alcohol (see methods of preservation, Appendix I). If they are to be observed further, they are transported to the laboratory in wide glass vessels lined with slightly moistened blotting paper or in small wire baskets.

In the quantitative investigation it is a question of finding out how many insects emerge in a certain time over a certain area of the stream. For this Mundie (1956) has constructed a special collecting funnel. This is a three-sided pyramid with a border 60 cm long. Two sides of it are covered with plexiglass (perspex) 3 mm thick, while the third side is covered with copper wire gauze (with 22 meshes/cm^2). The funnel stands on three legs, which should lift it 8 cm above the bottom of the stream, so that they must be about 12 cm long. It is braced with angle irons, which make it quite heavy and sufficiently strong. The covering of plexiglass is replaced at the tip with nylon gauze (22 meshes/cm^2) in order to prevent the deposition of condensation water (Fig. 87a). A wide-necked bottle on the tip of the funnel collects the insects that emerge, and a smaller funnel projecting into it would prevent them from getting out again; one can easily make such a small funnel oneself.

The collecting funnel is driven into the bottom in such a way that the sides covered with plexiglass are directed against the current (Fig. 87b). The side covered with gauze is directed downstream. Between the bottom of the stream and the lower border of the funnel there is an open interval of about 8 cm. The angles of the funnel are next attached to large stones in the river with stainless (rust-free) wire and, finally, also moored to the shore (Fig. 87b). This ensures that the funnel remains on the spot even at high water. Recently Mundie (1964) has described an improved method.

The insects which emerge while the funnel is exposed collect in the gauze or in the bottle above. The number of the animals of each species emerging is stated per unit of surface area and time.

A simpler container, covered with nylon gauze, which is driven into the bottom of a stream, has been used by Ide (1940), Sprules (1947), and Gledhill (1960) to determine the rate of emergence of stream insects. This container, closed on all sides, has a door through

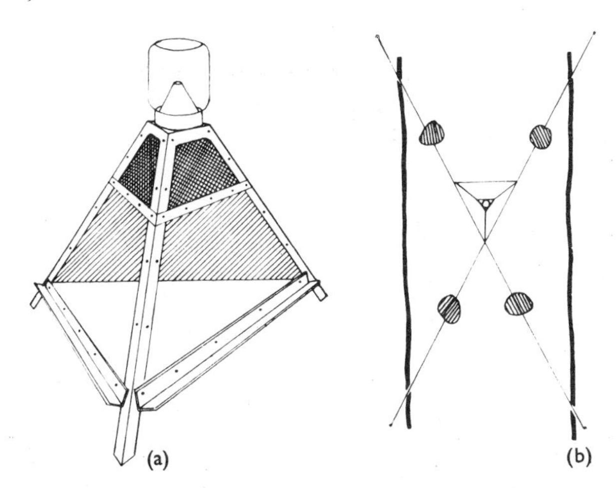

FIG. 87. Funnel for trapping insects emerging in running water: (a) the funnel itself; (b) its position and attachment in the stream.

which the insects emerging are collected. The device has the great disadvantage that the insects emerging are not retained, as they are by Mundie's pyramidal trap, but fall after a while into the water and are washed away; this kind of trap must therefore be checked relatively often.

Another simple trap has been used by Sommermann, Saller, and Esselbaugh (1955) to study the population dynamics of *Simulium* (black flies). Mundie (1956) has summarized the methods.

VI. SEMI-EXPERIMENTAL METHODS FOR THE DETERMINATION OF THE POPULATION OF RUNNING WATER

Important supplementary methods for the estimation of the quantitative population of running water are those by which a specific substrate is exposed in the water-body and whose colonization after a certain time is studied quantitatively. The study of the algal population, which is particularly difficult in running water, is especially made easier by the exposure of plates or films of this kind. In the interpretation of the results the position of the substrate exposed, its mineral composition, and its roughness must be borne in mind; all these factors may influence the colonization. For the study of colonization by animals the exposure of stones is useful—but these must naturally be free from organisms—and of clumps of plants, for which the same is true. By variation of the period of exposure one can investigate the progressive colonization of such substrates.

The corresponding methods for standing water are discussed in detail on pp. 81–84. Sládečková has also considered running water in her summary of the literature.

Exposure of plates and films for the determination of the algal population

Backhaus (1967) has very thoroughly investigated the algal colonization on artificial substrates exposed in flowing waters in the Black Forest and in the uppermost reaches of the Danube and has obtained very interesting results.† The best substrates proved to be roughened glass plates and plastic films, and it is still better to use only plastic films, which can be cut up as desired and examined under the microscope. It is best to mount the films on bricks, which are sufficiently heavy and also stay on the bottom in the more rapid currents. They are, however, often enough washed away at very high water. Figure 88 shows how the films are attached to the stones.

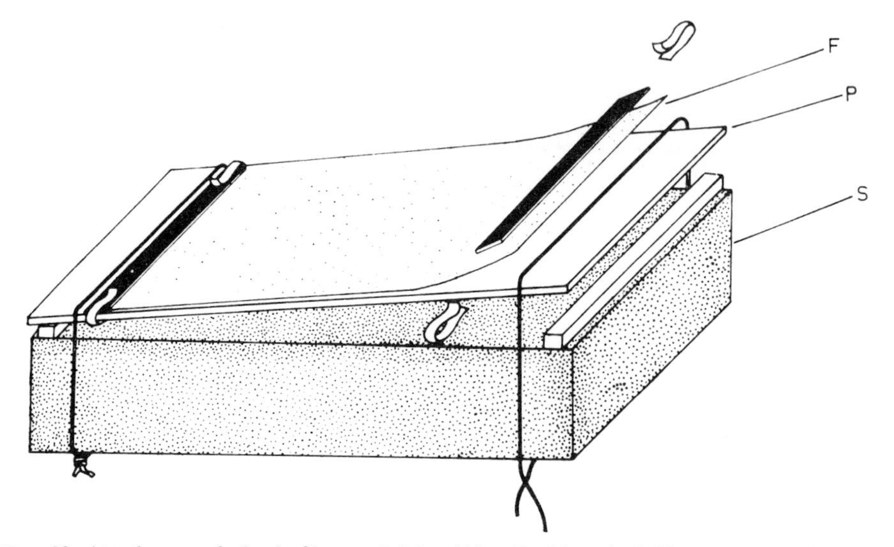

FIG. 88. Attachment of plastic films to bricks. (After Backhaus.) *F*, film; *P*, supporting plate (e.g. Eternit); *S*, brick.

The period of exposure depends on the problem in hand at the time. If one wishes to determine the sequence of colonization, one exposes several films and investigates one of these every 2 days. For the study of the colonization as a whole up to its final stage, an exposure for 30 days is, according to Backhaus, sufficient. The stones are then brought up, the films are detached, and the growth is estimated.

Films exposed for more than 30 days are mostly overgrown with several layers. This causes difficulties in the quantitative estimation because it is impossible to recognize the cells in layers several millimetres thick or to define or count them at all. For the determination of the species and their frequency individual coatings of the layer of growth, consisting chiefly of diatoms, and metaphyton‡ and detritus, must be scraped off with a bristle brush or a sharp scalpel and investigated separately. Then the true *Aufwuchs* species remain and these can then be determined. For the estimation of the frequency of the species the seven steps

† I thank Dr. Backhaus for his permission to anticipate here the methodological results.

‡ The metaphyton consists, according to Behre, (1956, 1958), of the group of free-living algae (unicellular and cell colonies) "which interpolate themselves between the algae of the growth and between the leaves of the submerged plants and into the pads of filamentous algae freely suspended in the water, without being firmly attached here and also without being plankters in the open water to any special extent".

of the scale (see p. 151) are used. The plate method is less suitable if one wishes to investigate the effect of the current on the density and speed of the colonization on substrates, as for this the plate is not sufficiently discriminative. One uses either plastic bands wound round a stake driven into the bottom so that one can thus also expose the bands at different depths; or the larger stones, for example, are completely covered with plastic bands which can then be later carefully taken out and investigated, as Backhaus has done. In all cases it is important for the later estimation to know accurately the relations of the current to each individual part of the films or bands.

Determination of the animal colonization

Stones and clumps of plants come under consideration as natural substrates, which must have been most carefully freed of all organisms before they are exposed. These substrates are investigated for their animal colonization after different periods of exposure in the water-bodies. It is best to use bricks or other slabs of stone as artificial substrates, but with these one should take into consideration the fact that a substrate that is foreign to water may possibly have an inhibitory effect on the colonization.

These investigations provide an insight into the rate of the colonization of substrates under different conditions (type of water, current, kind of the substrate, time of the year). Work of this kind is urgently desirable in the limnology of running water.

VII. FURTHER TREATMENT OF THE SAMPLES

This is done in the same way that samples taken from the littoral and bottom of standing water are treated. The sieving of the samples of sediment, and the selection of organisms and washing them clean, has been discussed in the section on Profundal (pp. 95 ff.) with particulars of various procedures.

Reference may here be made to the sorting apparatus (sorting trough) of Moon (1935) which, like Löffler's elutriating process (see pp. 96 ff.) sorts the sediment into different fractions. One can easily make the apparatus oneself (Fig. 89). It is a box about 130 by

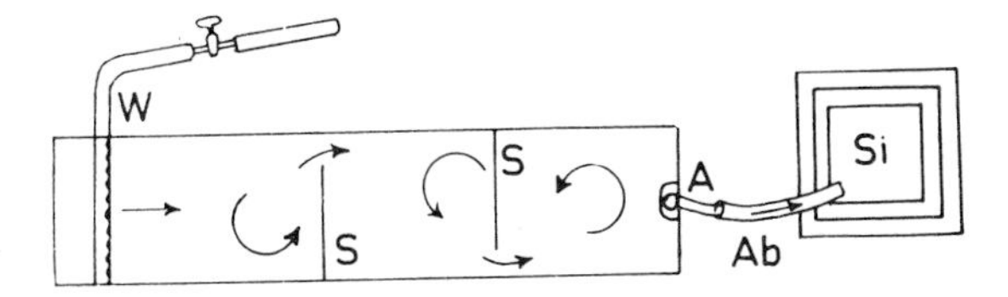

FIG. 89. Sorting apparatus (sorting trough) of Moon. (Redrawn from Albrecht, 1959.) (See text.)

20 by 6·4 cm which is divided into three compartments by two partititions which do not completely shut them off from one another. The sediment with the organisms is delivered into the first chamber, and water is led in through a lead pipe perforated with a series of holes. The stream of water forces itself through the slits left open by the incomplete partitions and flows from one chamber to the others, by which process the sediment is vigorously stirred up and is carried along for variable distances according to its weight. The lightest particles and most of the organisms come out of the container on to sieves of different mesh-widths. If the sample is only a small one, the partititions are not necessary and the light material can be washed directly in to the sieve.

Counting under the binocular

If the sample contains numerous very small organisms and a good deal of detritus as may, for example, be the case in hyporheic investigations, it is advisable to count only part of the sample under the binocular. To do this the whole sample is diluted a little, stirred up and 2 cm³ of it are put with a pipette into a Petri dish 7–8 cm in diameter, on the bottom of which parallel Indian ink lines have been painted about 5 mm apart. Under the binocular the space between these lines can be examined and the whole dish can be covered in the same way. The animals are thus counted. It is best first to look through some samples and to identify the groups or species present. Then lists are made out on which, when the accurate count is made, each group or species can be ticked off. In samples rich in organisms it is sufficient to count 5%, but in samples in which they are less numerous more must be counted or, in extreme cases, the whole sample. The numbers obtained in the counted cubic centimetre are converted to the whole sample (see Schwoerbel, 1961).

The statement of the production is made in grams of fresh or dry weight per unit of surface and, when it is necessary and possible, also per unit of substrate, of, for example, stones.

The frequency of individual species can also be estimated according to the seven-step scale of frequency, but in this there is a great risk of subjective errors of estimation according to the previous adaptation of the investigator.

VIII. PRESENTATION OF THE RESULTS

Some methods for the evaluation and presentation of biological results in the limnology of running water have come into use, and here they must be summarized briefly.

Estimation of the impurity of water according to biological conditions

This extremely important sphere of hydrobiology is discussed in detail on pp. 147 ff.

Biological zonation of running water

As was explained on p. 105, the conditions of life in running water, such as the temperature, current, and nature of the bottom of the stream, alter in the course of the river, and therefore the composition of the living community also changes. To represent this change in the living community along the course of a river, Illies (1953) has worked out a very instructive graphic representation of it.

The sites of investigation from the source of the river downstream are entered on the abscissa of a system of coordinates. The number of species obtained of one or more groups of organisms is entered on the ordinate at each site of investigation, and for the sites sampled above and below these an entry is made in each instance on the ordinate of how many of the species occur there. A system of curves results which includes as many curves as there are sites sampled. The maximum of each curve lies at the respective sampling site and the course and steepness of the curves show which sampling sites show a similar and a different population. From such a representation one can readily detect the effects of pollution as well as the changes in the fauna in sections of divergent hydrographic character, for example, small lakes (Fig. 90) traversed by rivers. Schmitz has treated these curves mathematically; Illies and Botosaneanu (1963) have given a survey of the literature.

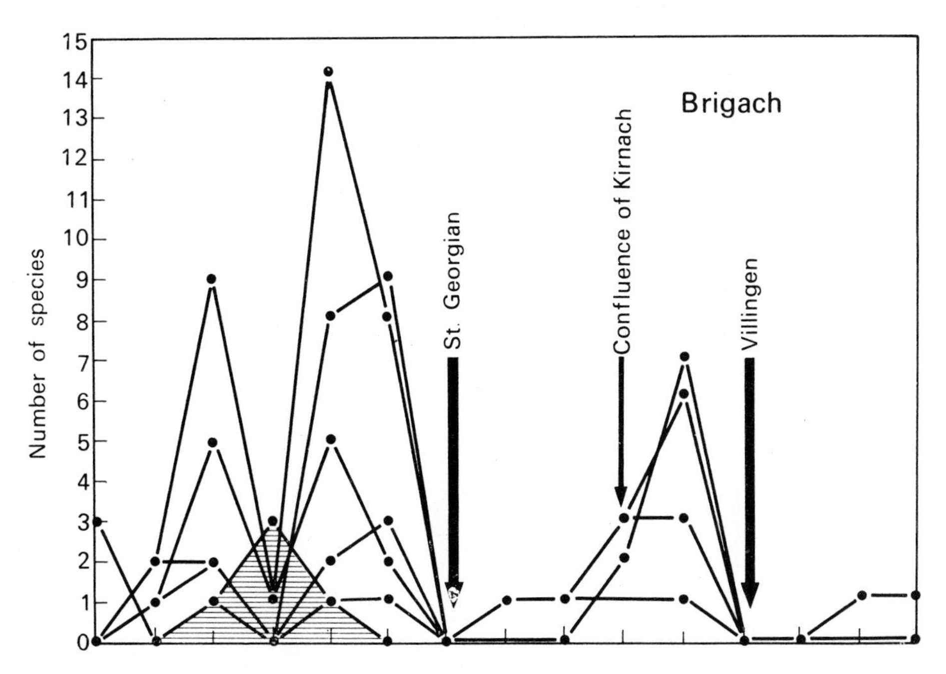

FIG. 90. Curves of the species of water-mites in the Brigach. (From Schwoerbel, 1964; see text.) The peak of the hatched-in area shows the number of species in a small pool which is interpolated into the course of the stream; these species are absent above and below.

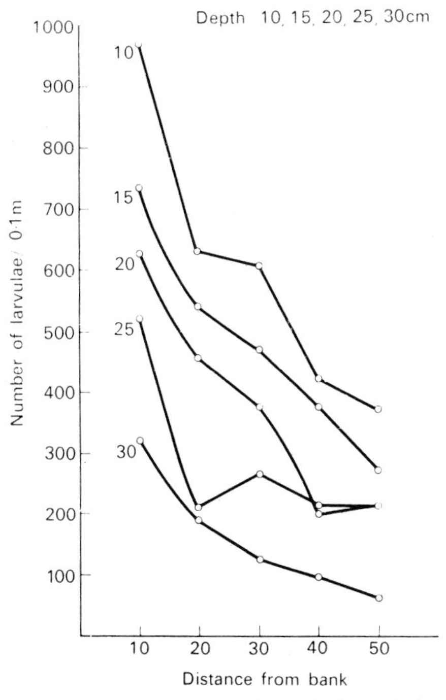

FIG. 91. Distribution of „larvulae" and chironomids in the hyporheic region; parameter—depth below the bed of the stream. (From Schwoerbel, 1964.)

Representation of the hyporheic distribution of organisms

In correspondence with the biological interchange in the river, the animal community in the hyporheic region changes with increasing depth and also with increasing distance from the water. In longitudinal sections and depth profiles one can show the quantitative displacement for the individual groups of organisms or for the individual species of a group (Schwoerbel, 1961); the representation of the absolute number of individuals per square metre (ordinate) at different distances from the stream (abscissa), with the respective depths as a parameter, provides a very informative picture of the distribution (Fig. 91).

CHAPTER 5

METHODS FOR THE BIOLOGICAL INVESTIGATION OF UNDERGROUND WATER

ACCORDING to Thienemann's (1925) definition the term underground water includes all the water that circulates in the outermost layers of the earth below its surface. It fills spaces and fissures in rocks, large chambers in the rocks, and also the small and the finest pores in the deposits of gravel and sand in depressions in river valleys and on the shores of lakes and seas.

This ground water is inhabited by animals which show interesting features according to the peculiarities of this sphere of life. Truly underground spheres of life are always devoid of light, and many inhabitants of this biotope have either reduced eyes or none at all. Also their activity lacks the characteristic day–night rhythm of animals living above ground. Their sense of touch is highly developed and is specialized by the development of many tactile hairs or very long antennae. The exclusion of the subterranean biotope from the climatic variations of the surface of the earth explains the constant low temperature of the underground water (in so far as it does not come from the greater depths). The truly subterranean organisms therefore lack the developmental rhythm that is evident in most of the animals that live above ground.

In caves the aquatic animals live in small collections of water, streams, and pools of water impregnated with mineral matter (sinter water), and also in the film of water on stalactites and stalagmites and on the walls of caves. In the stream of ground water of porous rocks, the so-called interstitial ground water, the environment is often extremely limited. The typical "small space inhabitants" are, without exception, very small and often very much elongated. Their limbs and antennae are reduced, and their movements are sinuous and jerky or crawling.

Underground water is accessible to investigation in caves or in porous rocks where special procedures are needed to open it up. Springs are natural outlets of ground water, and the so-called hyporheic region can also be regarded as a ground-water environment which, because of its importance for running water, has already been discussed in detail under that heading (p. 118). On the shores of lakes (hygropsammon) and the coasts of seas (mesopsammon) this interstitial pore water is also studied in the same way as in the hyporheic region. We deal here only with the water in caves and with the true ground water.

The organisms which live exclusively in caves are called troglobionts; examples of them are *Proteus anguineus*, a number of fish and higher Crustacea (Decapoda). Species which live in the cave spaces of broken stone, gravels, and sands (porous rocks), are stygobionts. They live in deposits in depressions in river valleys (nappe phréatique of Daubrée), so that they are called phreatobionts. Phreatic organisms which live directly between the river deposits under the bed of the stream and directly on the shores of brooks, in the region where the river and the ground water mix, are hyporheobiontic, the others being phreatobiontic in the strict sense. Those organisms which do not occur exclusively in the environments mentioned are *-phil* if they are numerous and occur regularly, and *-xen*, if it can be assumed that they penetrate only accidentally into the environments under investigation.

No special methods are needed for the study of collections of water in caves. On the other hand, "frequenting" caves presupposes special equipment, rubber clothing, carbide lamps, climbing irons, rope ladders, and a rubber dinghy. Naturally we cannot here go into the general methodology for the investigation of caves. For small pools and shallow "lakes" in caves it is sufficient to have a hand net (see Fig. 48) and a stock net made of silk gauze No. 20, because many organisms in cave waters, such as Crustacea and Halacaridae, are very small.

It is always advisable first to pass a light over the shore of the pool with a flashlight. In this way the larger animals can be identified and collected with forceps, a pipette, or carefully with a net. One should always utilize to the full the possibility of collecting animals in this way because they are very delicate and are easily pulverized or damaged when they are collected and separated from a sandy sediment. The following groups of animals are especially found in standing cave waters: planarians, copepods, and especially Harpacticidae, Amphipoda, and Isopoda, and some Decapoda and fishes. The dredge (see p. 78) can also be used, and also planktonic copepods, rotifers, etc. can be collected with the plankton net.

Animals living in cave waters can also be collected with the help of bait. Spandl (1926) has described a useful method, the bait-can. This consists of a metal can made of brass or aluminium, provided with a lid that closes tightly. The casing, bottom and lid of it are perforated by numerous holes 3–5 mm in diameter. Soldered on to the casing are two wire or tin rings. The can is loaded with a bait, the best one being raw meat or a big dead snail, and then it is anchored to stones under the water with two wires. In running water Turbellaria and Hirudinea (leeches) are lured to it and captured. In standing water the can is put in a plankton net which, when the can is pulled up, filters out the water flowing from the can and also the animals in it. Also pieces of cloth, simply wetted with water and put down at the edge of the water, always contain after a while numerous animals (Spandl, 1926).

Racovitza and Chappuis (1927) used, instead of a bait-can, a bait-net. It is a simple plankton net, over the openings of which two strings were stretched cross-wise; where they cross one another a small piece of liver or spleen is attached and the net is let down on a line weighted with a stone. Every 3–12 hr the net is pulled up and the animals captured are collected. These two authors also used with success a small cage with bait in it.

In streams in caves the methods used for running water above ground are employed (see Stammer, 1932), i.e. stones are turned round and searched, samples of sediment are sieved, and even hyporheic investigations can yield interesting results.

The smallest collections of water, such as the walls of caves irrigated with water, dripping rocks (stalactites) and their drop-filled basins on the bottom, also contain organisms which must be collected directly with forceps or a pipette.

It is self-evident that, in addition to the biological investigation, the temperature, pH, and hardness of the water should be determined, and, in addition, the nature of the bottom should be investigated.

Literature: Spandl, 1926; Chappuis, 1927; Leruth, 1939; Jeannel, 1926; Hamann, 1896.

INVESTIGATION OF GROUND WATER IN POROUS ROCKS (PHREATIC FAUNA)

The ground water which fills the pore spaces of porous rocks must be opened up by artificial means. As is known, this is achieved by wells and tanks which are unfortunately today even less common. However, a few decades ago Lais was able to collect the ground water fauna of the upper Rhine region by filtering the water of the wells provided with pumps in the villages, through fine plankton nets (see Kiefer, 1957). About 300 l. of water should be filtered from each well. In doing this it is useful to immerse the net in a bucket of water so that the water-pressure does not act directly on the animals captured and destroy them (Husmann, 1956). Nowadays there are hardly any wells with pumps. Chappuis discovered in the same way a large number of ground-water animals which were, until then, unknown, in the water-pipes of the Speleological Institute in Cluj (Romania). Tanks and the uncovered shafts of wells, which serve for the conservation of water for individual houses in mountainous regions, have also become rarer. The investigation of such tanks is done with a hand net or a net mounted on a stick. The detritus brought up is carefully washed out and put into a dish from which as many organisms as possible are selected alive.

Literature: Chappuis, 1922; Haine, 1946; Löffler, 1960; Noll and Stammer, 1953; Vejdovski, 1882; Jakobi, 1954.

One can also make pits on the shores of open water and thus obtain ground water, as Chappuis (1942) has done in running water and Pennak and others have done in lakes. Here the biotope of the boundary line between the surface and ground water is encountered with a characteristic mixed fauna. Such biotopes are the psammon of the shores of seas and lakes (see Littoral section, p. 71) and also the hyporheic sphere of life, already described

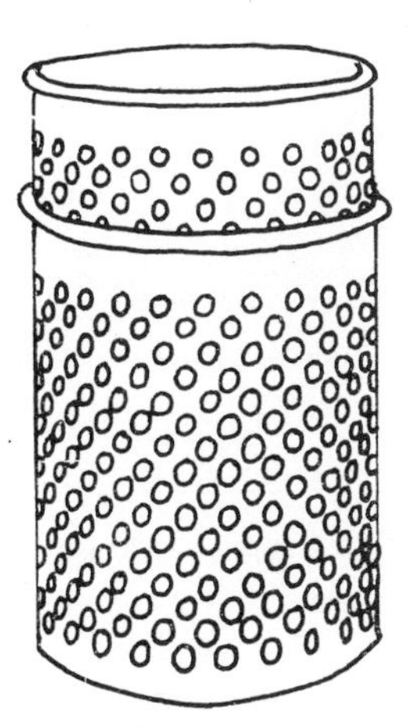

FIG. 92. Husmann's metal cylinder for diggings on the shore.

in detail (p. 118), which has been investigated especially by Orghidan (1959), Ruffo (1961) and Schwoerbel (1961), as well as by Angelier (1953). The method of using pits has already been described (p. 118). In order to avoid the troublesome later breakdown of the walls of the hole dug out, Husmann (1956) used a sieve-like perforated telescopic metal cylinder which is, during the excavation, driven down into the bottom of the hole being dug out (Fig. 92).

Interesting biological results can be obtained if one makes the pits at different depths and at different distances from the shore and relates the organisms caught to their dependence on the depth and distance from the shore (see Fig. 91). With regard to this see Schwoerbel (1961, 1961a).

Literature with information about methods: the shore ground water of running waters; Motas, 1962, 1962a, 1963; Motas, Tanasachi, and Orghidan 1957; Serban, 1963; Ruttner-Kolisko, 1961. Shore ground water of lakes: Sassuchin, 1931; Wiszniewski, 1934, 1935; Pennak, 1940, 1950; Ruttner-Kolisko, 1953, 1954, 1956. Marine shore ground water; Remane and Schulz, 1934.

In springs the ground water appears on the surface and organisms also are washed up with it. To collect these the simple funnel-collector of Noll (1939) (Fig. 93) is used. A double

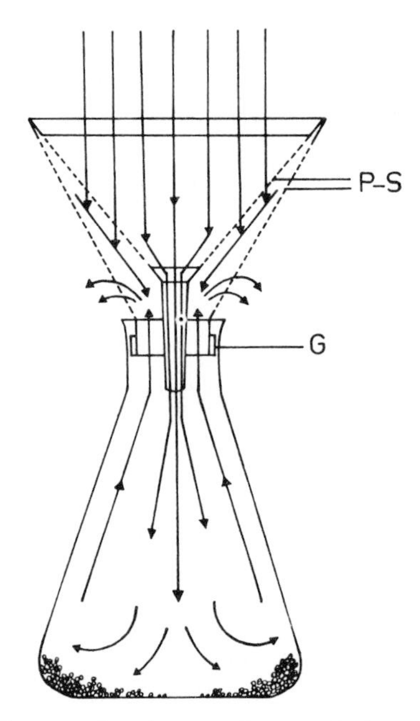

FIG. 93. Collecting funnel for ground water organisms washed out of springs. *P-S*, bronze-wire gauze; *G*, rubber gasket. The arrows show the flow of the water. (Redrawn from Noll and Stammer.)

funnel made of bronze wire gauze with the mesh-width of an ordinary plankton net is put on to a large thick-walled Erlenmeyer (conical) flask with a capacity of 2–5 l. The inner funnel projects for about 5 cm into a 2-l. flask; the outer funnel is fixed to the neck of the flask with a rubber gasket. It is best to bury the apparatus overnight in the outlet of the spring and to examine it in the morning. Detritus and organisms are washed into the flask and collect on the bottom, and the filtered water leaves it through the outer funnel of the apparatus. The contents of the flask are not immediately preserved, but the animals are taken out alive.

Husmann (1956) has improved the apparatus; in his "filter glass" the jet of water is first collected by a funnel and then led through a rubber tube into the nozzle of the inlet tube (Fig. 94b). This tube also passes through the middle of the circular perforated cover-plate and also through the solid bottom plate of a hollow cylinder made of phosphor-bronze gauze. The gauze cylinder is fitted into a glass cylinder with a diameter about 1·5 cm larger, by means of a rubber gasket. The water drawn in can only escape through the openings in the cover-plate after passing through the gauze and particles of detritus are thus filtered out (Fig. 94a). To collect water from the outlets of springs one can attach to the apparatus, instead of the funnel, a small metal collector, in the manner shown in Fig. 94b.

By the methods mentioned, organisms can be collected from all the spheres of life in underground water. The sorting out of samples rich in detritus is done in Petri dishes under a binocular lens as described on p. 126 for the hyporheic fauna. The organisms are collected alive and then are fixed as usual (see Appendix I). For the most modern descriptions of the ground-water fauna we may name here the books of Delamare-Deboutteville (1961) and Vandel (1964).

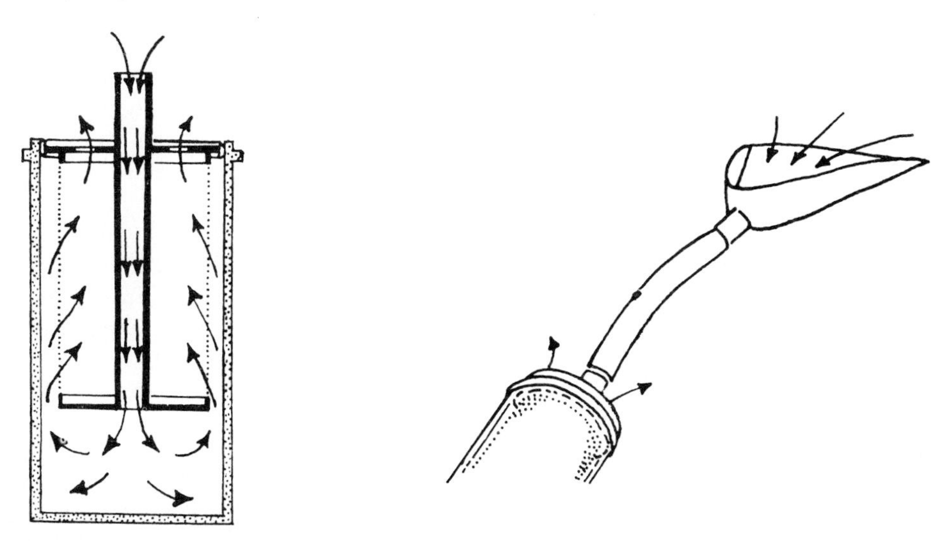

FIG. 94. Husmann's filter glass for ground-water animals in springs: (a) the filter-glass, (b) inlet tube with collecting funnel. (Redrawn from Husmann.)

CHAPTER 6

METHODS FOR THE DETERMINATION OF PRODUCTION IN WATER

MANY critical considerations and investigations have been devoted to production in water. Recently Elster (1954) and Davis (1963) have expressed opinions on it, and Ohle (1956), Rodhe (1961), and Elster (1963) have considered it especially from the point of view of metabolism dynamics.

The classical definition of the concept production was given by Thienemann (1931): "Production of a biotope in organic substance within a given time is the total quantity of organisms and their excreta formed within the biotope during this time."

This definition is no longer valid today. Strictly interpreted the production of a water-body can be a question only of that organic substance which is formed by photo-autotrophic and chemo-autotrophic plants out of the inorganic nutrients in the water and the energy irradiated into the water from without. This "primary production" is accomplished in the first place in the trophogenic layer of the chlorophyll-bearing plankton and littoral plants. The extent of the chemosynthetic primary production in the tropholytic zone is still largely unknown.

This primary production is the basis of the whole biogenic metabolic cycle in waters; *all the remainder* is consumption and decay. The animal consumers live on the primary production; next come the plant-feeders themselves (primary consumers), then the predators (secondary consumers), and, finally, through more or less numerous intermediate stages, the fishes. Therefore it should be strictly noted that, so far as the animal consumers are concerned, it is a question not of a true production but only of the reconstruction of organic substances already present.

This reconstruction results in the formation of animal proteins and other structural materials, and it also provides the motive energy for the functioning of the living system. Stated in terms of energy, the consumption is therefore always associated with a loss of energy, and production is associated with a gain of energy.

If, therefore, one speaks for example of production in zooplankton or fish, "production" is meant only in this figurative sense. This "production" is always smaller than in the overlying levels of "food and being food", so that the graphic representation of a trophic pyramid results, the base of which is the primary production and the apex is the yield of fish (Fig. 95).

This trophic pyramid is, however, only a part of the biogenic metabolic cycle. The cycle is completed by the chemical and bacterial decomposition of the organisms which begins immediately after death; the nutrient substances go back again more or less quickly into the water and are again available for the construction of organic substance (see Krause, 1964). Numerous recent studies have shown that the cycle can be short-circuited completely in this way.

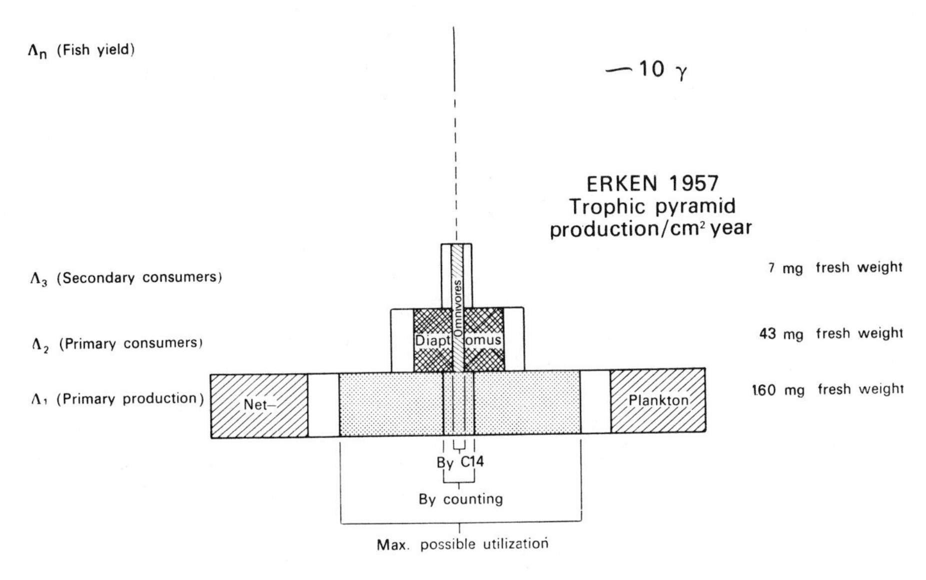

FIG. 95. Trophic pyramid of the successive levels of the synthesis (production) and reconstruction (consumption) of organic substance. (From Nauwerck, 1963.)

For the determination of the production, the estimation of the biomass, i.e. the mass of organic substance momentarily present (standing crop), is important. Production is then the renewal of this biomass in the unit of time (= turn-over). Increase and loss are separate items in the turn-over. The following statement then applies:

$$\text{Biomass}_2 = \text{Biomass}_1 + \text{increase} - \text{loss}.$$

The biomass and the increase or loss are determined. The investigation of production is done in limnology in so many different ways that only some basic indications, related exclusively to primary production and primary consumption, can be given in the present description.

I. DETERMINATION OF THE BIOMASS

As can be seen from the trophic pyramid, the determination of the biomass of the phytoplankton, related to a unit of surface or volume of lake, is a completely adequate basis for the estimation of the primary production.

Determination of the biomass is, however, valid physiologically and ecologically only if it takes into consideration the proportion of individual species in the biomass. To determine this, water-bottle samples are taken from the water-body, as has been described in the section in phytoplankton, and these are fixed with Lugol's solution and counted for algae under the inverted microscope. This gives the number of species and individuals of primary producers per unit of volume or beneath 1 m² of lake surface.

For the calculation of the actual biomass, however, the volume of the individual primary producers must be borne in mind, the determination of which has been described on p. 7. But it may be noted in relation to this that the volume of individual species may vary widely from lake to lake. For this reason the values obtained by Nauwerck (1963) are not reproduced.

Three additional methods are mentioned below which provide an average of the respective biomass without entering into the proportion of the individual species. Thereby a part from each of the zooplankton, bacteria, and organic detritus is also dealt with.

1. Chlorophyll Methods

According to the plankton content, 0·5–5 l. of water are filtered under reduced pressure through a "coarse Cellafilter". The residue in the filter, which contains bacteria and detritus as well as phyto-plankton, is kept in steam for 45 sec to break down the chlorophyllases, is then lightly dried, and then brought into 90% acetone for 20 hr and put in the dark. Šesták (1958) breaks down the cells before the extraction with quartz sand. To estimate the chlorophyll the extinction of the acetone extract is determined in a spectrophotometer at wavelength of 665 m, 645 m, and 630 m (colour filter!). These wavelengths correspond to the absorption maximum for chlorophyll a, b, and c. Kalle (1955), on the other hand, uses a colorimeter method which uses the fluorescent light of chlorophyll and also much smaller quantities of water because even weak solutions of chlorophyll are strongly fluorescent.

To calculate the individual chlorophyll constituents of a sample, the extinction values of the individual fractions are multiplied by the factors stated (according to Parsons and Strickland, 1963, and Richards and Thomson, 1952), and subtracted from one another according to the following scheme:

$$\text{Chlorophyll a} = 11{\cdot}6\ D_{665}\ (15{\cdot}6) - 0{\cdot}14\ D_{630}\ (0{\cdot}8) - 1{\cdot}31\ D_{645}\ (2{\cdot}0)$$

$$\text{Chlorophyll b} = 20{\cdot}7\ D_{645}\ (25{\cdot}4) - 4{\cdot}34\ D_{665}\ (4{\cdot}4) - 4{\cdot}42\ D_{630}\ (10{\cdot}3)$$

$$\text{Chlorophyll c} = 55\ \ D_{630}\ (109) - 16{\cdot}3\ D_{645}\ (28{\cdot}7) - 4{\cdot}46\ D_{665}\ (12{\cdot}5)$$

These factors are valid for 1 cm cuvettes.

The final calculation of the chlorophyll content of a sample under investigation is given by the formula:

$$\text{mg chlorophyll (a, b, or c)/m}^3 = \frac{C(\text{a, b or c})v}{lV},$$

in which v is the amount of acetone extract in ml, V is the volume of filtered water, and l is the length of the cuvette used to determine the extinction.

The method naturally provides only conclusions about the phytoplankton. The values so far found lie between 10 and 120 μg of chlorophyll per litre (Gessner, 1959).

Literature: Richards and Thompson, 1952; Gessner, 1959; Parsons and Strickland, 1963; Aruga, 1966.

2. The Dry Weight Determination Method

Birge and Juday (1922), in order to compare the productivity of individual lakes, first used estimations of dry weight. In order to obtain the zooplankton and phytoplankton, large quantities of water (more than 1 m^3) must be filtered through a plankton net and then centrifuged. The residue in both the net and after centrifugation is collected, dried, and weighed. The dried substance is then related to a unit of surface or volume. In Wisconsin lakes the organic dried substance, which naturally also contained organic detritus and bacteria, varied according to the time of year between 0·23 and 12 mg per litre and per hectare, excluding the shore region, between 258 and 522 kg.

3. The Nitrogen Method

The total nitrogen in a sample of water with a natural content of plankton is estimated before and after filtration through a membrane filter. The difference between the two estimations gives the nitrogen content of the suspension, further including the inanimate particles and the zooplankton and the bacterial plankton. Because protein contains on an average 16% of nitrogen, the nitrogen value obtained multiplied by 6·25 gives the respective quantity in crude (gross) protein. The weight of the total organic dry substance is obtained approximately by multiplication of the nitrogen value by 20.

II. DETERMINATION OF THE PRIMARY PRODUCTION

The primary production of photo-autotrophic plants is determined by estimation of their photosynthetic capacity. To do this, basically three methods are available: the oxygen method, the C^{14} method, and the chlorophyll method which has already been partially described.

1. The Oxygen Method (Gran Method)

According to the equation given on p. 6 the amount of carbon dioxide absorbed in carbon assimilation is proportional to the oxygen liberated by plants so that the amount of carbon assimilated can be calculated from the amount of oxygen.

Procedure

Lake or river water is taken with a plexiglass (perspex, lucite) water-bottle from a certain depth and put—free of air bubbles—into two bottles with a capacity of 100–200 ml. One bottle remains clear, the other is given a coat of black paint or is enclosed in a light-proof box (dark bottle). Both bottles are then let down into the water, i.e. to the depth from which the water was taken. This operation must be done as quickly as possible. During the exposure of the bottles the oxygen content of the water used is determined by the Winkler method (procedure on p. 23). After the period of exposure, which varies according to the plankton content (from 6 to 24 hr), the bottles are taken out of the water and their oxygen content is determined. From the initial value obtained previously and the oxygen content of the clear bottle, the amount of photosynthetically formed oxygen is obtained from which the photosynthetic capacity of the plants can be calculated. The darkened bottle serves as a control, and give the oxygen consumption in the respiration and also the bacterial oxygen consumption. The amounts of assimilated carbon determined by several experiments or series of experiments are appropriately given in grams of carbon per cubic metre or per volume of water beneath 1 m^2 of water surface, because it is also certainly proportional to the irradiated energy of the surface.

Sources of error

1. There is a risk that the algae in the closed flasks sediment down during the exposure, so that the photosynthetic capacity is thus impaired. The smaller the flasks are, the less is the sedimentation. Backhaus (1967) has worked in mountain streams with test-tubes suspended in such a way that they were kept in constant movement by the water current (see Fig. 97). This made sedimentation or the formation of envelopes of water supersaturated with oxygen around the algae impossible.

2. When the plankton content was small it was possible to take small flasks and expose them longer. However, the bacteria multiplied very rapidly on the inner glass walls during the period of exposure so that increased consumption of oxygen resulted. There is also a light-dependant inhibition of bacterial growth by the algae so that, after longer periods of exposure, the dark flasks no longer provide comparable control values.

3. Influences emanating from the glass of the flasks used must be considered, so that the same flasks must always be used.

4. Naturally air bubbles in the flasks must be avoided; therefore immerse the filled flasks again immediately to the depth from which the sample is being taken or provide in experimental investigations for a suitable uniform and constant temperature.

5. The method is not suitable for vascular plants, which accumulate larger amounts of oxygen in their parenchyma (Wetzel, 1965).

2. The C^{14} Method of Steemann Nielsen (1952)†

The principle of this method is the following. The photosynthetic capacity of a phytoplankton population is measured by the amount of fixed carbon. When the investigation is set up the total amount of carbon available is marked with a trace of radioactive C^{14}. After the experimental period the uptake of C^{14} can be related to the total uptake of carbon during this period. Even very small amounts of C^{14} can be measured by the Geiger–Müller counter. Because of its greater mass C^{14} is, however, assimilated about 6% less than C^{12} (Steemann Nielsen, 1955). Strickland (1960) has subjected the method to a general critical examination, but many methodological problems are still not explained.

Mode of working

1. Prepare the experiment.
2. Draw off a water sample to determine the plankton content.
3. Estimate the total carbon in the water under investigation.
4. Treatment of the experimental bottles after their exposure.
5. Determine the impulse count of the fixed C^{14} in the Geiger–Müller counter.
6. Calculation of the amount of fixed carbon.

With regard to the preparation of the experiment, making the C^{14} solution, distribution in and standardization of the ampoules, see pp. 143 ff.

Procedure for the determination

The work is done in the sequence described above.

1. The experiment is prepared in the following stages: taking the water sample, filling the experimental bottles, addition of the C^{14}, and exposure of the bottles in the waterbody.

To take the water sample, a plexiglass (perspex) sampling bottle is used, not a metal one. Two bottles of about 100 ml capacity are filled, quite free from air bubbles.

When glass vessels are used, one should bear in mind the fact that these vessels have, at different lake depths, a variable transparency to ultraviolet light (Findenegg), so that the

† The method involves so many difficulties and modifications that in this description of it only the method of working and the most importand sources of error can be discussed. A symposium held in March 1965 in the Institute of Hydrobiology in Pallanza on Lake Maggiore was chiefly concerned with primary production and the C^{14} method. Goldman and Vollenweider, *Methods for measuring primary productivity in aquatic environments*, University of California Press. Berkeley, 1966.

comparison of the series of measurements is made difficult. In relation to bottle-glass —arbitrarily 100—the transparencies to ultraviolet light for Jena glass and quartz glass are:

0·0 m deep	Jena glass 85	Quartz glass 57
0·2 m deep	Jena glass 95	Quartz glass 60
0·1 m deep	Jena glass 107	Quartz glass 95

To each flask is added 1 ml of a standard solution of a bicarbonate $NaHC^{14}O_3$ of accurately known radioactivity (see p. 143). The strength of the dose of C^{14} is determined by the plankton content of the water. Of the two flasks one is darkened. Both flasks are immersed to the depth at which the water sample was taken. During their exposure the flasks should have a horizontal, not a perpendicular, position. In this way the surface illuminated is greater, better use is made of the light, and the stoppers of the flasks do not create shadows (Elster and Motsch, 1966). To avoid a *lightshock* in the algae the bottles must be again immersed in the lake as quickly as possible.

About the period of exposure there is no uniformity. Rodhe (1958) and Nauwerck (1963) exposed them in the Swedish Lake Erken for 24 hr, Elster and his collaborators (1963) exposed them for 5 hr in south German lakes. This interval of time might be more favourable.

2. *Emptying out a water sample to determine its plankton content.* About 100 ml of the bottle sample are put into a tightly screw-topped bottle. Lugol's solution with added acetate is added to it, and the sample is put aside for later counting of the plankton. For the method, see pp. 59 ff.

3. *Determination of the total carbon in the water under investigation.* This proceeds in three stages: determination of the pH value, the alkalinity, and abstracting the carbon value from the table of Harvey and Rodhe.

The pH is estimated either colorimetrically or electrochemically, but it must be accurate to 0·1 units. To determine the alkalinity, 100 ml of the water sample are titrated with N/10 HCl and methyl orange as an indicator (see p. 30). From Table 6 by Harvey and Rodhe,

TABLE 6. CALCULATION OF THE TOTAL CARBON IN WATER FROM THE pH AND ALKALINITY (DATA FROM BUCH AND HARVEY FROM RODHE)

pH	f_C	f_C in free CO_2	pH	f_C	f_C in free CO_2
6·0	42·14	30·29	7·6	12·73	0·759
6·1	35·95	23·96	7·7	12·58	0·600
6·2	31·02	19·05	7·8	12·43	0·475
6·3	27·11	15·12	7·9	12·34	0·374
6·4	24·00	12·01	8·0	12·25	0·300
6·5	21·53	9·80	8·1	12·17	0·237
6·6	19·57	7·69	8·2	12·11	0·187
6·7	18·01	6·03	8·3	12·05	0·149
6·8	16·76	4·78	8·4	12·03	0·119
6·9	15·79	3·79	8·5	11·94	0·0928
7·0	15·01	3·03	8·6	11·88	0·0729
7·1	14·39	2·40	8·7	11·82	0·0576
7·2	13·90	1·91	8·8	11·74	0·0453
7·3	13·50	1·51	8·9	11·66	—
7·4	13·19	1·20	9·0	11·56	—
7·5	12·94	0·953			

the total carbon can be determined from the basis of the alkalinity value for 100 ml of water.

The value ascertained for the alkalinity (100 ml of water) is multiplied by the factor f_C corresponding to the pH of the water (total carbon in F) or f_{CO_2} (total carbon in free CO_2).

4. *Treatment of the experimental bottles after their exposure.* After the period of exposure the bottles are taken out of the lake and their contents are filtered and dried.

The clear bottles are darkened immediately they are taken out of the lake. As quickly as possible the contents of the bottle, or if the plankton is dense, a part of the contents (25 ml) are vacuum filtered (in a suction filter) through a membrane filter No. 2 with a pore diameter of 400 m. Wash again the residue with distilled water or 0·005 N HCl in order to dissolve out adsorbed inorganic carbonate.

The filter with the filter residue thus purified is dried in a desiccator for at least 24 hr. The dried filter can be stored for a long time (the half-life period of carbon is a round 5570 years) and can be sent away in suitable packing.

5. *Determination of the impulse count of the "filtered residue" in the Geiger–Müller counter.* Care should be taken in doing this that the layer on the filter is not too thick because then one must reckon with a self-absorption of the impulse within the algal layer and the values obtained are too low. Therefore when there is vigorous phytoplankton development only a part of the sample should be filtered the volume of which must, however, be accurately known.

C^{14} is a β-emitter and decays into N^{14}; so that under the Geiger–Müller counter the β-activity of the dried algal residue is measured. The amount of assimilated C^{14} can be estimated from the impulse number per unit of time, and from that the total assimilated carbon can be calculated.

A *windowless methane gas-flow* counter is inserted as a counter. In modern instruments the samples are automatically put on a "roundabout" in the counting device, and the impulse number per unit of time is recorded.

6. *Calculation of the amount of carbon taken up.* The following proportionality equation is valid according to the conditions of the experiment:

$$\frac{\text{Fixed impulses}}{\text{Total impulses}} = \frac{\text{Fixed C}}{\text{Total C}},$$

from which it follows that

$$\text{Fixed C} = \frac{\text{Fixed impulses} \times \text{Total C}}{\text{Total impulses}}$$

The calculation of the assimilated carbon in milligrams per hour and cubic metres of water volume then follows from the following calculation:

$$\text{mg fixed C/h m}^3 = \frac{\Delta\ \text{Impulse}\ h-d \times \text{Total C} \times \text{bottle vol.} \times 1{,}000}{\text{Total impulses} \times \text{Filtrate vol.} \times \text{Time in hours}}$$

Total carbon is obtained from Table 6 and the total impulse number from the standard value of the ampoule (see p. 144); Δ Impulse $h-d$ is the impulse difference between the clear and dark bottles.

Some sources of error

The method starts from the assumption that no carbon is lost through respiration that is certainly produced simultaneously with photosynthesis and by excretion. This is not in any way so, and the respiratory loss is considerable during long-lasting experiments. Because, in this case, the carbon liberated in respiration originates from photosynthesis, one measures by the C^{14} method only the net production, and therefore the absolute gain of substance; this may be the advantage of an exposure time of 24 hr. In an interesting way there is also a fixation of carbon in the dark bottles which should in general amount to 1–2% of the values in the clear bottles (Gessner, 1959), but is higher in impure waters with a high bacterial content. A high carbon assimilation value in clear and dark bottles has also been found in the deep zone of lakes, which is certainly to be ascribed to a carbon fixation by chemo-autotrophic algae and bacteria. Investigations into this subject are urgently necessary.

Goldman's method

Goldman used another method to determine the activity of the fixed algal material He burnt the material in a closed system and discharged the gas thorugh the counting chamber of the counter. This procedure measures the absolute C^{14} concentration and operates largely without loss. A standardization is not necessary here as it is in the filter-counting method. Goldman obtained, in comparison with the filter-count values, substantially higher activity values; they were especially high with diatoms whose cell walls have obviously a very high specific absorption for the radiation from the carbon isotope; in other algae this self-absorption is lower. A decisive advantage of the gas-phase method is that these self-absorptions are eliminated completely; and another advantage is that the method is generally substantially more sensitive.

Use of the method and expression of the results

Because the photosynthetic capacity of the algal population depends, apart from its density, on other factors such as the light intensity and temperature, experiments are made *in situ* at different depths of a lake. Thus several pairs of bottles placed one below the other are exposed in vertical series so that they do not cast shadows on each other. Naturally, in order to obtain an accurate evaluation of the experiment, the illumination at each of the depths investigated during the period of exposure must be measured and preferably also the colour constituents. These measurements are nowadays most accurately made with a photometer with a hemispherical receiving surface. Likewise an accurate estimate of the intensity of light at the surface of the water is necessary.

The vertical series *in situ* mostly show a characteristic maximum of the production capacity only under the surface of the water and not perhaps at the surface itself (Fig. 96), because at the surface the intensity of light is too strong and inhibits photosynthesis and the ultraviolet light irradiated in has a destructive effect on the chlorophyll. In the individual lakes and types of lakes the curves of a vertical series have, however, quite different courses (Findenegg, 1964). As the temperature rises the optimal light intensity for photosynthesis also rises (Talling, 1957).

Use of the C^{14} method in mountain streams

The C^{14} method may also provide good results with phytoplankton in slow-running water-bodies. In small streams, in which the primary production is accomplished by the

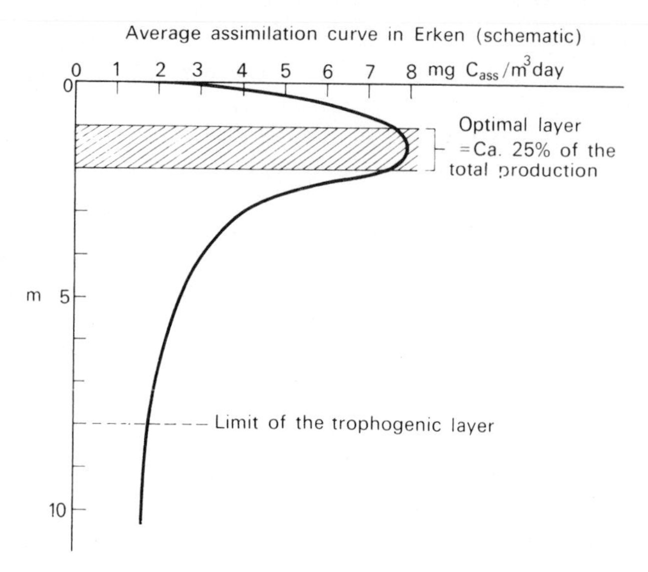

FIG. 96. Course of a photosynthesis curve of lake phytoplankton often observed, obtained by the C^{14} method in Lake Erken. (From Nauwerck, 1963.)

water-plants rooted in the bottom and by the *Aufwuchs* algae, Backhaus (1967) used the method rather differently.

Plastic films were exposed (for precise details about these see p. 124) for a certain time in the water, during which time a coating of *Aufwuchs* algae became attached to the films. The films were taken out and two small discs were cut out of them with a tubular punch, these being exactly of such a size that they fit into the capsule envelope of a Geiger counter. The clear and dark vessels each receive one of the small discs. These vessels were test-tubes 15 cm long with a diameter of 3 cm. The dark tube was coated with black varnish and covered with insulating tape to protect it from scratches.

The test-tubes were filled with water from the stream, the carbon content of which was calculated from the pH (see Table 6), alkalinity, and with 1 ml of C^{14} solution, a small air bubble being left in each tube. A thin plastic pin about the size of a match-stick attached to the cork inside the tube prevents the plastic disc from sticking to the underside of the stopper.

The clear and dark tubes were fixed in the stream in such a way that they swing in the current (Fig. 97), and the air bubble then keeps the water in the tube itself moving. Backhaus exposed them for 4 hr and then took out the discs and carefully washed them with distilled water. The discs were then put into the counting capsule with the algal layer side uppermost, and dried by the standard method in a desiccator.

For very accurate work, the water in the clear and dark vessels must also be filtered and separately measured because algae may have been detached from the films during the exposure time.

If the algal growth is very thick, its self-absorption may influence the results. Backhaus therefore detached a part of the layer before it was dried, dried this in another capsule, and counted it separately. When the growth on the films is less, two or more discs can be put into the experimental vessels.

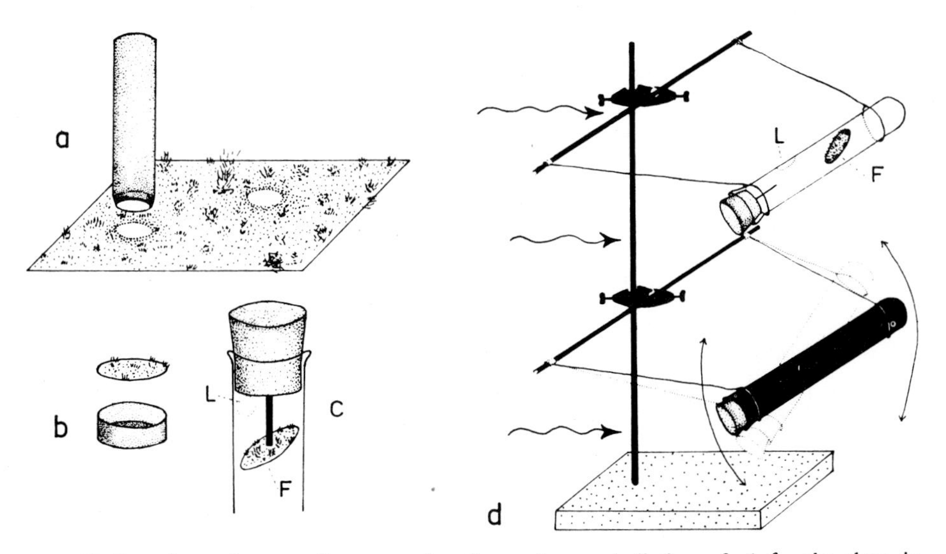

FIG. 97. Experimental set-up for measuring the carbon assimilation of *Aufwuchs* algae in running water (after Backhaus): *a*, punch and plastic film covered with growth; *b*, counting cell and plastic disc suitable for the sample changer of the Geiger–Müller counter; *c*, rubber stopper with plastic pin; *d*, stand with experimental vessels in the current; *F*, small plastic disc; *L*, air bubble.

The calculation proceeds in the manner described above, thus:

$$C_{\text{fixed}} = \frac{\text{Impulse } h-d \times C_{\text{total}}}{\text{Impulse}_{\text{total}}}.$$

With a population density of 100 cm² and an exposure time of 1 hr, it follows that

$$C_{\text{fixed}}/100\ \text{cm}^2/\text{h} = \frac{\Delta\,\text{Impulse } h-d \times C_{\text{total}} \times 100}{\text{Impulse}_{\text{total}} \times F \times h},$$

in which C_{fixed} is the amount of carbon in milligrams taken up by the algae during the period of the experiment; C_{total} is the total amount of carbon in milligrams in the experimental vessels at the beginning of the experiment; $\text{Impulse}_{\text{total}}$ is the initial total impulse number in the experiment vessels (=ampoule value); Δ Impulse $h-d$ is the difference between the impulse values in the dark and clear vessels; and F is the area of the discs.

Preparing the C^{14} *standard solution from* $BaC^{14}O_3$

In an evacuated apparatus (Fig. 98) the $BaC^{14}O_3$ is slowly treated with dilute hydrochloric acid and the $C^{14}O_2$ produced is absorbed on to sodium hydroxide exposed to it for 1 hr. The following reactions occur:

$$BaC^{14}O_3 + 2\,HCl = BaCl_2 + H_2O + C^{14}O_2$$

$$C^{14}O_2 + NaOH = NaHC^{14}O_3$$

The sodium hydroxide containing $NaHC^{14}O_3$ is diluted until it has the theoretically desired activity. The pH of the diluted solutions is adjusted to 9·5 (displacement of the carbonate

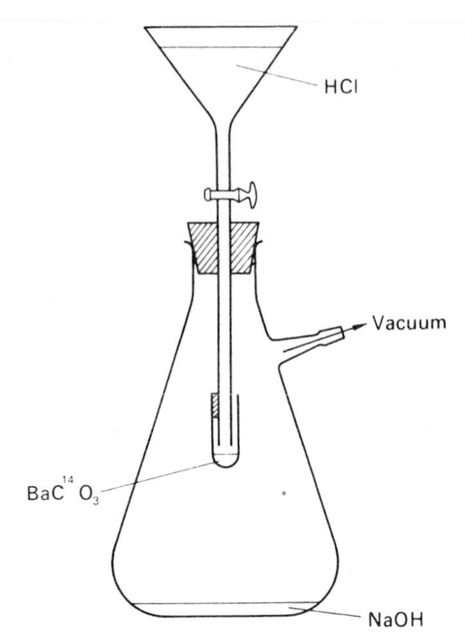

FIG. 98. Apparatus for preparing the C^{14} standard solution (see text).

equilibrium to HCO_3) and the solution is put into sterile ampoules. The ampoules from a stock of which (e.g. 1000 of them) supplies are drawn, are closed, sterilized in the autoclave, and kept cool and in the dark. Because of the long half-life of carbon, a loss of activity need not be feared.

To destroy the residues the radioactive carbonate is precipitated as $BaC^{14}O_3$ and the waste is subsequently sent to the authorities or to the institute (e.g. Radiological Institute). Work with radioactive substances requires special authority, and protective measures laid down by the law are necessary.

Standardization of the ampoules to determine the total impulse number

A weighed amount of 53·7 mg of previously dried sodium carbonate is dissolved in some distilled water and put into a 100 ml measuring flask. The solution is made alkaline with 1 ml of concentrated ammonia and the content of one ampoule (1 ml) is added. 2 ml of 10% $BaCl_2$ solution are added as a precipitant; the precipitated, non-radioactive barium carbonate serves only as a carrier substance for the radioactive carbonate. The small measuring flask is filled up to 100 ml with distilled water. The solution thus made contains per 100 ml per 100 mg of $BaCO_3+BaC^{14}O_3$ as a precipitate.

From the well-shaken solution, 0·25 ml, 0·50 ml, 0·75 ml, 1·0 ml, 2·0 ml, 3·0 ml, 4·0 ml, and 5·0 ml are successively filtered off by suction filtration on to a membrane filter. The precipitate should be distributed as uniformly as possible over the filter.

The samples are dried in a desiccator and counted in a Geiger–Müller counter. The impulse numbers for the different amounts of fluid are set out in a diagram, the abscissae of which show milligrams of barium carbonate and the ordinates the impulse numbers. The curve thus obtained runs linear only within the range of 0·25–1 mg of $BaCO_3$. When there are larger amounts of barium carbonate precipitate on the filter, the self-absorption of the β-radiation in the precipitate becomes evident. From the linear representation, however, the total impulse number for 100 mg $BaCO_3 + BaC^{14}O_3$ = 1 ml of the standard solution can be obtained by rectilinear extrapolation.

Three ampoules from each series of them, i.e. about 1000 of them, are tested in the manner described above.

In order to keep errors in the making of the standard solution as small as possible, it is more advantageous to work with a larger weighed amount. If one takes a weighed amount of 268·5 mg of previously dried sodium carbonate, and fills up with 250 ml of distilled water, then 50 ml of this solution contain 53·7 mg of Na_2CO_3.

III. SOME REMARKS ABOUT THE DETERMINATION OF THE PRIMARY CONSUMPTION

All the organisms which have no means of carrying out photo- or chemosynthesis, feed either directly or through intermediate stages on the products of the assimilation of green plants; they constitute the group of consumers. The bacteria also belong to this group; during rapid destruction they nevertheless take up nutrients although still in the trophogenic zone, and may pass them on to the zooplankton. These short-circuited cycles play a decisive part in lakes.

The relationships between the primary producers and the consumers and also those between the individual consumers are diverse, i.e. it has so far been possible to follow them out only in isolated series. This work has been done for practical reasons especially on the nutrition of fish and so-called "food chains" have been established. The precise nutritional requirements per unit time of only a few species of zooplankton have so far been investigated, e.g. those of *Daphnia pulex* (Richmann, 1958) and *Leptodora kindtii* (Sebestyén, 1959).

The consumption of the primary production begins with the filter-feeders (primary consumers) and is continued by the predators (secondary consumers). Filter-feeders and predators are, however, not restricted to these two groups of consumers, but there are also all intermediate stages, and some species may be filter-feeders as well as predators.

Recently an elegant method has been worked out by which the filtration rate of the filtering primary consumers can be determined experimentally. This method is closely connected with the C^{14} method described: by it an accurately known concentration of algae made radioactive by assimilation of C^{14}, i.e. "marked", is fed to zooplankton, and the quantity of water filtered free from algae (the filtration rate) is calculated from the quantity of algae taken up per unit of time (the feeding rate), this latter being calculated from the radioactivity of the animals.

Procedure for determination of the filtration rate

Lake water containing its natural plankton population (bottle sampler) is filtered through bolting cloth No. 25 so that all the zooplankton and net phyto-plankton is removed. To it is added $Na_2C^{14}O_3$ solution, as in the C^{14} method and the water sample, in which only the nannoplankton still remains, is exposed in vessels for 24–48 hr under optimal light conditions and at a temperature corresponding to that at which it was taken. After this period of time the algae have taken up sufficient C^{14} and can be used for the determination of the filtration rate.

About 50 ml of the radioactive sample is filtered, dried, and counted under a Geiger counter (initial zero value); another sample is fixed with Lugol's solution.

The animals to be investigated must be taken fresh from the lake (by net collection). They are taken out individually from the catch, washed in sample water and distilled water, and then put with a suspension of algae into accurately standardized experimental vessels of 10–125 ml capacity. The number of animals necessary for an experiment depends on their size; thus 10–20 specimens of *Diaptomus* would be used and 30–40 specimens of *Bosmina*. One experimental vessel remains without any animals, and shows at the end of the investigation the changes in the water samples that have occurred independently of the influence of the animals (final blank sample).

The duration of exposure, during which the animals can ingest radioactive algae, should be not longer than 0·5–4 hr because the investigations must be concluded before the animals excrete any of the algae taken in.

To determine the food taken up, the animals are removed from the experimental vessels and washed with distilled water, dried on a gauze filter in a desiccator, and, finally, their radioactivity is measured by a Geiger counter. Likewise, in addition, the final blank sample and its radioactivity is worked out.

The average value between the initial and final blank samples gives the average radioactivity of the food material during the experiment; it is stated in impulse number per millilitre. The filtration rate of each animal is then obtained from the average blank value of the food and the radioactivity of the animal:

$$\text{Filtration rate} = \frac{\text{Impulses per animal}}{\text{Impulses per millilitre}}$$

Nauwerck found for *Eudiaptomus graciloides* a filtration rate of 0·3–2·8 ml per animal per day and for *Daphnia longispina hyalina* 0·2–4·5 ml per animal per day. This value corresponded well with those obtained by Richman for *D. pulex*, i.e. 0·2 to over 5 ml, and that obtained by Rigler, i.e. 2·5 ml per animal per day, for *D. magna*. Marshall and Orr (1955) used P^{32}-marked algae for similar investigations on *Calanus*.

The filtration rate depends on various external factors, e.g. on the temperature and the food supply. It reaches, however, even under the most favourable conditions, a maximal effect which cannot be exceeded. Then the amount of food taken in depends solely on the concentration of the food. Nauwerck found, on the basis of his experiments and investigations in Lake Erken, that the zooplankters studied by him can take up only about 5% of the nannoplankton in the lake; to supply the energy requirements a filtration rate 10–100 times greater was necessary.

IV. POPULATION DYNAMICS OF THE ZOOPLANKTON

In order to estimate the secondary consumption, a knowledge of the changes in the stock of a zooplankton species is very informative. According to Elster (1954) a stock B_2 arises from a preceding stock B_1 plus growth minus losses:

$B_2 = B_1 + \text{growth}\,(Z) - \text{losses}\,(V)$.

B_1 and B_2 are estimated by means of estimations of the respective stocks (see zooplankton methods, pp. 34 ff.) that are done one after the other at short intervals of time. The growth Z is experimentally determined from the amount of eggs taken in a catch and the developmental period of the eggs at the temperatures in the lake (=the "renewal coefficient"). The expected stock B_2' is obtained from B_1 and the growth Z, and this is always higher than the actual stock B_2. $B_2' - B_2$ gives the loss figure.

It goes without saying that investigations of the population dynamics of individual species of zooplankton must be done separately and should extend over at least a year. Because no details can be given here, reference is made for more precise information to Elster (1954), Eichhorn (1957), Eckstein (1964), Nauwerck (1963), and also Edmondson (1960, 1962).

CHAPTER 7

METHODS FOR THE BIOLOGICAL ESTIMATION OF WATER QUALITY

THE rapid and reliable estimation of the degree of pollution of our waters that are charged with sewage is a practical necessity of some significance. Since the fundamental researches of Lauterborn (1901, 1903) and those of Kolkwitz and Marsson (1902, 1908, 1909), biological evaluation of it has accompanied chemical investigation, and in this way a more important advance in the evaluation of the condition of our polluted waters has been attained.

The biological evaluation of polluted water starts from the problem of how the biological situation in a drop of water is altered when waste water containing poisonous substances or organic substances that foul it are discharged into it. The organic substances are first broken down (mineralized) by decomposers. The intermediate and end products of this decomposition become available to the consumers and primary producers. The substance added to the water are thus incorporated by the agency of the organisms into the metabolic system of the water. This process, in which adsorption processes in the sediment also participate, is called, with regard to the situation initially prevailing, self-purification. Knöpp (1964) quite rightly differentiates between an oxidative or heterotrophic phase of the self-purification and a photosynthetic or autotrophic phase of it.

The process of self-purification is thus associated with organisms, being begun by the bacteria as decomposers and continued by the chlorophyll-bearing primary producers. The actual significance of the biological evaluation of water loaded with sewage therefore lies in both the determination of the load of sewage and in the biological capacity of the water for self-purification which cannot be correctly estimated by any chemical method. The actual load of sewage can be recorded chemically only in individual values obtained at the moment; the biological analysis, on the other hand, provides insight into lasting effects on the water.

I. ECOLOGICAL METHODS

In every sphere of life a change in the environmental conditions leads to a more or less radical transformation of the living communities. This is true of life on land as well as in water. In every piece of water into which sewage is discharged, a similar situation arises; the results depending on the characteristics of the sewage and the amount of it discharged. If the initial amount of it discharged is very large, the whole of the oxygen is used up to mineralize the organic substances. The lives of all organisms which need oxygen in order to live are then in the highest degree imperilled. The living community is altered in favour of those organisms which manage either without oxygen or with very little of it. In addition to this are permitted only those organisms which can live on organic nutrients. In the course of the self-purification, less oxygen is required for the mineralization and oxygen supply for the organisms again becomes more favourable. By this the extreme initial situation is altered again in favour of the organisms needing oxygen, among which initially only the heterotrophic species and later the autotrophic ones also can gain a footing. In the most favourable case the situation adjusts itself to what it was before the entry of the sewage (Fig. 99). The trophic grade of the water is, however, higher, because the potential in inorganic nutrients has been increased. This operates especially on the closed system of a standing piece of water (e.g. Lake Zürich and the Lake of Constance).

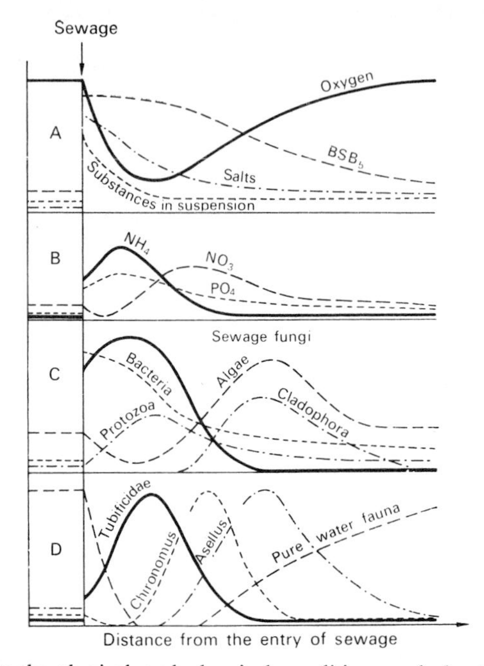

FIG. 99. Alterations in the physical and chemical conditions and also the colonization in the course of running water below the entry of organic waste water (sewage). (Redrawn and modified from Hynes.)

The biological–ecological estimation of water purity starts from the statement that the stress of impurity in each instance can be deduced from the biological situation without knowing the chemical sewage characteristics in detail. The assumption is that the biological situation is typical for each degree of pollution. The biological analysis is made along two lines—either some of the organisms typical of the degree of pollution, the so-called indicator organisms, are established, and their occurrence is followed, as is done by procedures which work with the saprobic system, or the impoverishment of a "natural" species spectrum is brought in as a basis for the evaluation, as Kothé has done quite logically with his "species deficit". Other procedures which determine the relationships of reducers, consumers, and producers to one another, refer exclusively to the process of self-purification.

1. Procedures which Operate According to Saprobic Systems

The system of Kolkwitz and Marsson

Cohn (1853) and his pupil Mez (1898) did the earliest researches on the biological assessment of water. Lauterborn (1901) originated the conception of the sapropelic world of life by which he meant the organisms of decaying mud at the bottom of standing waters, and consequently he defined the saprobic zone as the zone of processes of decomposition. Kolkwitz and Marsson (1902, 1908, 1909) and Kolkwitz (1950) established the first extensive saprobic system with a large number of indicator organisms for the individual zones of contamination. The fact that this system has been used with good results until today in the practice of sewage biology, in spite of much criticism and discussion, shows how correct this system was. Liebmann (1947, 1962) first revised it on the basis of new scientific knowledge, and has extended it also to standing waters for which it was originally not conceived (see below). Subsequently there was a very lively discussion of the basis of the system (for this, see Caspers and Schulz, 1960, 1962; Caspers, 1966; Knöpp, 1962; Elster 1962).

The system of Kolkwitz and Marsson comprises four saprobic levels:

Very heavily contaminated	the polysaprobic zone (*p*)
Heavily contaminated	α-mesosaprobic zone (α)
Moderately contaminated:	β-mesosaprobic zone (β)
Scarcely contaminated	oligosaprobic zone (*o*)

Each organism recorded in a saprobic system is allotted to one or two of these zones that are adjacent to one another. If one investigates a chosen section of a flowing waterbody, one can draw conclusions about the amount of sewage from the number and frequency of these indicator organisms. For the practical procedure, see below pp. 151 ff.

Liebman's revision

Liebmann retained the four classical zones, but designated them as classes of quality and named them, in the series described above, as classes of quality IV–I. He further recommended the adoption of the cosmopolitan Protozoa for the evaluation of water quality. By doing this, methodological difficulties in the identification of the organisms would, of course, arise; the detection of the end phases of self-purification would, moreover, not be biologically correct.

The system of Fjerdingstad (1964)

Fjerdingstad further subdivided the zones of the classical system and used particular communities of bacteria and algae as indicators (Table 7).

Fjerdingstad's system uses only a few indicator organisms for the delimitation of the nine zones. It is primarily a matter of a documentation of the self-purification process. In practice there is the difficulty that some algae, especially *Cladophora* and *Phormidium*, cannot be identified with certainty or can be identified only with great expenditure of time (by culture). When the system is used it must be remembered that it was created in relation to the conditions in Danish waters, i.e. generally in waters in level country. According to Backhaus's researches it is valid only in rivers rich in lime.

The system of Sládeček (1961)

Sládeček, and before him Šrámek-Hušek (1956), have extended the classical system in the direction of especially heavy impurities (eusaprobic conditions), also including toxic sewage and radioactive contaminations (transsaprobic conditions) (Table 8).

In this system all kinds and degrees of contamination are taken into consideration. Naturally the abiotic zone of waters cannot be characterized biologically.

TABLE 7.

Zone	Communities
I. Coprozoic zone	Community of bacteria or of flagellate *Bodoibal* or both.
II. α-polysaprobic zone	1. *Euglena* community. 2. Community of rhodo- and thio-bacteria
III. β-polysaprobic zone	1. *Beggiatoa* community 2. Community of *Thiothrix nivea* 3. *Euglena* community
IV. γ-polysaprobic zone	1. Community of *Oscillatoria chlorina* 2. Community of *Sphaerotilus natans*
V. α-mesosaprobic zone	Community of *Ulothrix zonata* (high trophy) or *Oscillatoria benthonicum* or *Stigeoclonium tenue*
VI. β-mesosaprobic	Community of *Cladophora fracta* or *Phormidium*
VII. γ-mesosaprobic zone	Community of Rhodophyceae *(Batrachospermum moniliforme* or *Lemanea fluviatilis)* or community of Chlorophyceae *(Cladophora glomerata* or *Ulothrix zonata)*
VIII. Oligosaprobic zone	Community of Chlorophyceae *(Draparnaldia glomerata)* or a pure community of *Meridion circulare* or community of Rhodophyceae *(Lemanea annulata, Batrachospermum vagum,* or *Hildebrandia rivularis)* or community of *Vaucheria sessilis* or *Phormidium inundatum*
IX. Katharobic zone	Community of Chlorophyceae *(Chlorotylium cataractum* and *Draparnaldia plumosa* or community of Rhodophyceae *(Chantransia chalybdea* and *Hildenbrandia rivularis* or community of encrusting algae *(Chamaesiphon polonicus* and various species of *Cladothrix)*

1, 2, and 3 are various degrees of impurity within a zone.

TABLE 8.

Katharobic conditions (very clean water)	Katharobic conditions	
Limnosaprobic conditions (contaminated surface and ground water)	Oligosaprobic β-mesosaprobic α-mesosaprobic Polysaprobic	Corresponding to the system of Kolkwitz and Marsson
Eusaprobic conditions (domestic and industrial waste water, the basis of bacterial destruction)	Isosaprobic Metasaprobic Hypersaprobic Ultrasaprobic	Zone of ciliates Zone of colourless flagellates Zone of bacteria Abiotic zone but still not toxic
Transsaprobic conditions (without bacterial breakdown)	Antisaprobic Radiosaprobic	Toxic zone Radioactive waste water

Arrangement of the work

The practical execution of biological estimation of water quality by the saprobic system proceeds by the following stages:

1. Collection of the organisms in a defined area.
2. Identification of the organisms and determination of their abundance.
3. Interpretation and statement of the results.

Collection of the organisms

This is discussed in detail in individual sections of this book so that remarks on it here are unnecessary. It is better to collect qualitatively in a larger area than to collect quantitatively in a smaller one; nevertheless, all the substrates must be considered. The sites at which samples are taken are defined at the beginning of the investigation. These must be sufficiently near to one another, i.e. all the physiological characteristics of the water must be considered, should be readily accessible at all water levels, and the substrate to be considered must be extensive enough to avoid the recording of isolated findings. It is also important that the sites under investigation should be studied at all times of the year and, if possible, repeatedly.

Identification and abundance of the organisms

For the determination of the animals and plants the books on identification given in the bibliography can be used, and also the work of Lund (1960). The books of Liebmann, the figures in which greatly facilitate the identification, are especially concerned with indicator organisms, as well as Sládeček's (1963) beautiful *Guide to Limnosaprobical Organisms*. The larger organisms especially can be identified on the spot and their abundance can be estimated; all the others are fixed and investigated in the laboratory.

The abundance of individual species at a site under investigation is stated either in absolute numbers of individuals (method of Zelinka and Marvan) or it is estimated. Knöpp (1955) used a scale of 7 degrees (stages): 1, a single finding; 2, few; 3, few to average numbers; 4, average numbers; 5, average to many; 6, many; 7, abundant. According to Knöpp and also Mauch (1963), one estimates the highest for any stage. If it is assumed that a stretch of water is investigated by only one, and always by the same, worker, the personal errors of estimation are the same and scarcely matter. Nevertheless, they may be significant in a comparison of different stretches of water with different workers. Pantle and Buck use only three stages of frequency: 1, chance findings; 3, numerous; 5, abundant; naturally these stages are more easily assessed.

Interpretation and statement of the results

Knöpp's method. Knöpp (1965) investigated the animal and plant population of the shore and bottom zone of a stretch of river and worked out a "biological longitudinal section of quality". In doing this, two factors must be borne in mind—the presence of indicator organisms and their abundance. At each site investigated the species found are allocated to the four degrees of saprobic conditions and their frequency at the sites investigated are stated in estimated values of 1–7. The frequency numbers of the species at each degree of saprobic conditions are added up, so that the numbers of species belonging to two adjacent degrees of saprobic conditions are distributed accordingly (analogously) (2 : 1 or otherwise, according to their abundance). The results of this evaluation are thus the four sums Σo, $\Sigma \beta$, $\Sigma \alpha$, and Σp (corresponding to the four degrees of saprobic conditions on p. 149).

F

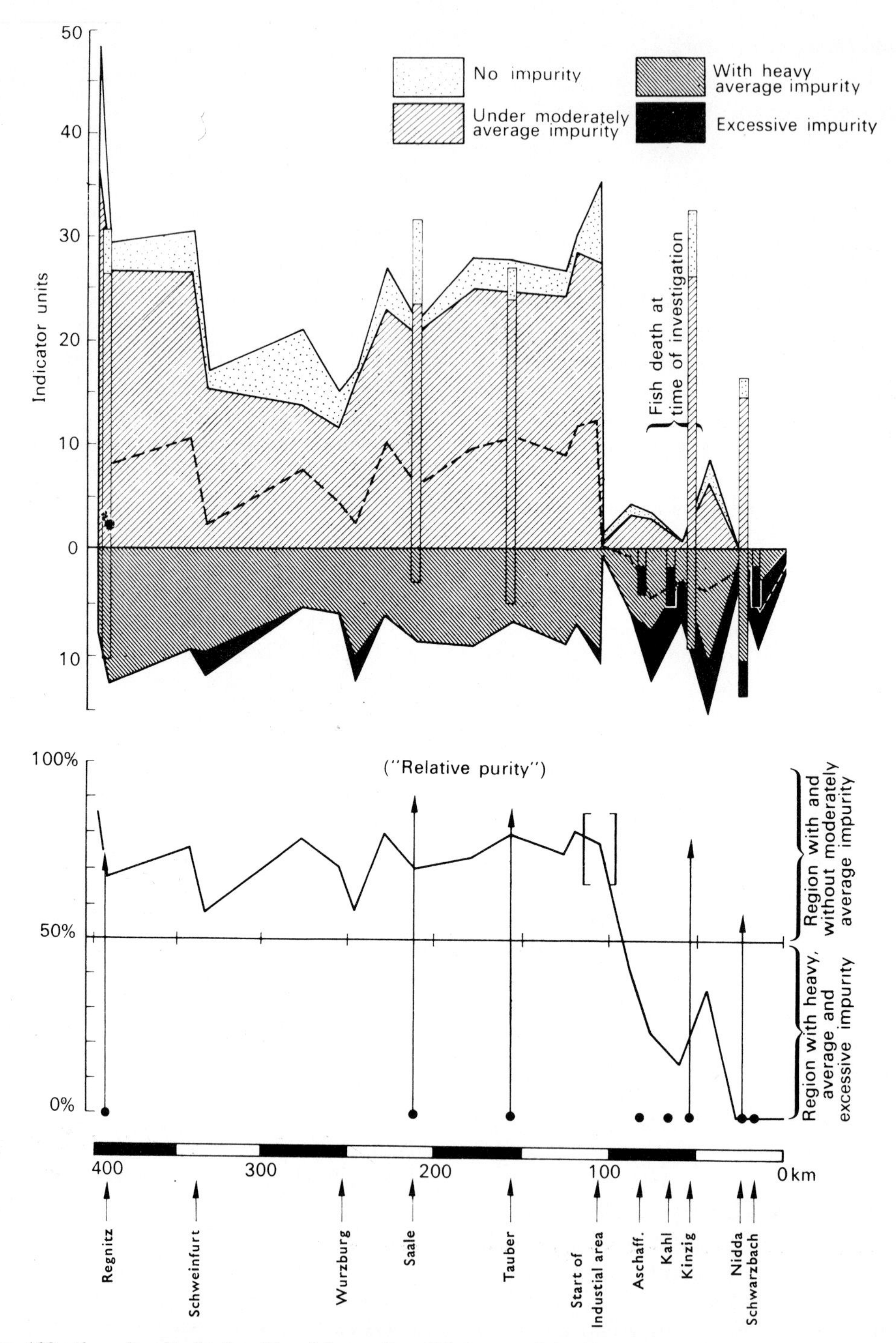

FIG. 100. *Above*: longitudinal section of the quality of the water of the River Main according Knöpp's graphic method. *Below:* representation of the relative purity of the same stretch of river. In both diagrams the points of the curves are joined to one another by straight lines. (Redrawn and somewhat modified from Knöpp, 1955.)

In the graphic representation of the results (biological longitudinal section of water quality) the four summation values are plotted out on an x-axis upwards or downwards. The x-axis represents the flow of the water so that the sites of investigation can be marked out true to scale. At each site of investigation the so-called positive values Σo and $\Sigma\beta$ are plotted above and the negative values $\Sigma\alpha$ and Σp are plotted below. The values of the same degrees of saprobic conditions are joined together and the resultant areas are emphasized by dots, cross-shading, etc. (Fig. 100 upper part). The biological longitudinal section of the water quality thus delineated shows very clearly the load (of impurity) at the individual sites of investigation and throughout the course of the river.

In addition Knöpp introduced the conceptions of "relative purity" and "relative loading (of impurity)", which can be calculated from the abundance summations of the individual indicator organisms:

$$\text{Relative purity} = \frac{\Sigma o+\beta}{\Sigma o+\beta+\alpha+p} \quad (\text{in}\,\%)$$

$$\text{Relative loading} = \frac{\Sigma p+\alpha}{\Sigma p+\beta+\alpha+p} \quad (\text{in}\,\%)$$

These values can also be represented very clearly by drawing a curve for the whole course of the river (Fig. 100, lower part).

Method of Pantle and Buck (1955). The frequency h of each species found which can be allotted to the classical system of saprobic conditions is estimated from the frequency scale of the three stages given above. Their saprobic grading s, and therefore their position in the saprobic system, is taken from Liebmann's (1952, 1962) lists: it is for

oligosaprobic indicator organisms	$s = 1$
β-mesosaprobic indicator organisms	$s = 2$
α-mesosaprobic indicator organisms	$s = 3$
polysaprobic indicator organisms	$s = 4$

From these statements the index of saprobic conditions, to be calculated for each site investigated, is

$$= \frac{\Sigma}{\Sigma h}$$

The purity is classified according to the following scheme:

= 1·0–1·5,	very slight impurity	(o)
1·5–2·5,	moderate impurity	(β)
2·5–3·5,	heavy impurity	(α)
3·5–4·0,	very heavy impurity	(p)

To make a visual illustration of the results, the values determined for s are plotted in different colours on the course of the stream (see also the Munich method, p. 155).

The procedure of Pantle and Buck was, like Knöpp's method, designed for running water. Schräder (1959) has, however, been able to determine also the degree of contamination of reservoirs with satisfactory results by this method. Breitig (1961) worked in running waters, and von Tümpling and Ziemann (1961) show that the degree of error for s is only small. Von Tümpling (1960) has shown by the example of the Werra stream that the curves of the saprobic indices correspond very well with Knöpp's relative loading.

Method of Zelinka and Marvan (1961). Zelinka and Marvan have investigated how a large number of organisms from standing and running waters are distributed in a spectrum of saprobic conditions consisting of five degrees. The five degrees correspond to the zones of the classical system, but the oligosaprobic degree is placed in front of an absolutely clean xenosaprobic zone. They determined for each species by their own investigations, by comparison with chemical data, and also according to statements in the literature, its distribution in the degrees of saprobic conditions, and marked this by numbers in such a way that the sum of this saprobial valency equals 10 for each species, as Table 9 shows.

TABLE 9.

Species	*x*	*o*	*β*	*α*	*p*	*g*
A_1	3	3	3	1		1
A_2	5	5				3
A_3	4	6				3
A_4	2	7	1			3
A_5	†	8	2			4

† indicates an isolated finding

This saprobial valency does not correspond to the frequency of the species in the degrees of saprobic conditions but is only an expression of a centre-of-gravity distribution in accordance with which each species is given an "indicator value" which characterizes its value as an indicator organism (see under *g* in the last column of Table 9). Species which in occurrence favour one degree of saprobic conditions (A_5) have a higher indication value than species which are diffusely distributed. The highest predicate is 5 and the lowest is 1. Beer (1958) and von Tümpling (1960) have also used earlier similar subdivisions.

The abundance of each species is calculated from the samples collected (individual numbers, no estimated ones), and this number is multiplied by the saprobial valency of the species concerned in each degree of saprobic conditions and also with its indicator value. The results of this calculation for the example selected are given in Table 10.

TABLE 10.

Species	Abundance (*h*)	Multiplication results for				
		x	*o*	*β*	*α*	*p*
A_1	69	207	207	207	69	
A_2	31	465	465			
A_3	30	360	540			
A_4	42	252	882	126		
A_5	8	+	256	64		
TOTAL		1284	2350	397	69	0
AVERAGE		3·13	5·73	0·97	0·17	0

The calculation of the degree of saprobic conditions is done by the following formula: for the xenosaprobic zone;

$$x = \frac{\Sigma xhg}{\Sigma hg}$$

for the oligosaprobic zone;

$$o = \frac{\Sigma ohg}{\Sigma hg}$$

and correspondingly for the *a* and *p* zones.

Thus for A_1 the abundance 69 is multiplied by 3 (= the saprobial valency in *x*) and 1 (= the indicator value) and the result 207 for *x* is recorded, and so on for each species and each degree. Then the individual values in each degree of saprobic conditions are added (Σxhg), i.e. for *x*, 1284, and divided by the sum of *h* times *g*. In our example this sum is 410. If one divides the totals of the individual degrees of saprobic conditions by 410, one gets for each of these degrees a "weighted" average value; the sum of all the average values again amount to 10.

Each site of investigation is analysed in the manner described. The position of the highest average value determines the final degree of saprobic conditions at each investigation site. The numerical relationship of the individual average values to one another can be stated graphically very clearly.

The method of Zelinka and Marvan is more time-consuming than the one previously described, but is more exact because its hypothetical basis is more correct. The saprobial valency is better characterized than the allotment to a single zone of the saprobic system; also the abundance of the species is not estimated but is counted. In particular cases one has to judge whether the gain in accuracy justifies the greater expenditure of time.

The statements about the saprobial valency and indication value of many species are readily available (Zelinka and Marvan, 1961). The procedure can be used for both running and standing waters. Sládečková and Sládeček (1963) have determined in this way the degree of contamination of dams, proceeding especially by determination of the *Aufwuchs* on exposed glass plates (for this method see p. 81).

Method of Liebmann (Munich procedure). The Munich procedure is based on the classical saprobic system which Liebmann derived from the evaluation of standing waters. In contrast to the procedures so far mentioned, the Munich school rejects all mathematical formulation and relies exclusively on the personal experience of the investigator. Fundamentally each organism is considered, and the experts know what value it has for the evaluation of these waters. Experience of this kind is also the basis of the tables of Zelinka and Marvan, but here they seem to be objectively fixed; without this basis the result varies according to the investigator's experience.

The Munich procedure records the classes of purity in the outline or the course of the water-body investigated. It thus makes a chart and picture of the quality of the water. Colours are used for the four classes of quality, blue for class I, green for class II, yellow for class III, and red for class IV; the intermediate stages are correspondingly divided (but the colours are not mixed). The resultant charts (e.g. for Lake Constance, the Danube, the Main) are very instructive, and show in the alarming yellow and red colours the danger points of the contamination of the water. We cannot here enter into the discussion caused by the application of this procedure to standing water or the more recent investigations into this (e.g. Zahner, 1964).

Sources of error in methods using saprobic systems

The most important objection that has been made to the saprobic system maintains that the organisms recorded in them have not been studied physiologically, so that we do not know what information they give as indicator organisms. This is true; in a very few cases we have accurate information, and the classical system depends on the personal experience of Kolkwitz, Marsson, Lauterborn, and others who have collaborated. In practice what is needed is a method for the biological evaluation of waters, and all those who use the system work in the same way and with their personal experience. This has always proved successful and, with regard to it, hydrobiologists are intensely concerned about an accurate knowledge of the biological requirements of the indicator organisms; not in order to reject the system finally, but in order to find out why it is so useful, Ambühl has especially occupied himself with the significance of the water current, and Zimmermann has investigated semiexperimentally the influence of the temperature and current on the living community of a polluted water in artificial channels. This has indeed shown that the estimation of the degree of pollution on the basis of indicator organisms depends very greatly on these two factors, because the indicator organisms especially react to them. The same pollution is more favourably estimated in rapidly flowing water and at lower temperatures than it is in slowly running water at a higher temperature (about this, see pp. 7 ff. and 104). This is a weighty objection to the use of indicator organisms whose biological requirements are not accurately known. In addition the tables of Zelinka and Marvan are strictly valid only for the waters in which they were worked out.

2. The "Species Deficit" of Kothé

Because of the difficulties just indicated, Kothé (1962) gave up entirely the saprobic system and the indicator organisms and used only the species density which decreases in an astonishing manner under the influence of both organic and toxic impurities in water. An advantage is that fundamentally all the species of plants and animals participate in this criterion, and that it is not necessary to allot them to particular degrees of contamination.

In practice a species standard is laid down which indicates a uniform standard of reference for all sites investigated. This species standard is obtained at an investigation site above the contaminated stretch of water, usually at the starting point of the whole investigation (sampling site 1).

Only the total number of the species is important; what species they are is immaterial. Assuming that each sampling site is investigated in the same way, the "species deficit" can be calculated from the following formula:

$$F = \frac{A_1 - A_x}{A_1} = 100,$$

in which A_x is the number of species at an investigation site x chosen at random and A_1 is the number of species at the sampling site 1, serving as the basis of reference. The results are stated in percentage values which vary between $0(A_x - A_1$ = no species deficit) and 100% (complete absence of species in A_x); in addition the species standard used must be quoted.

The values are clearly recorded graphically in a coordinate system, the abscissa of which shows the river stretch. It is advisable to combine the method with another method of evaluation; in this way there is good correspondence with Knöpp's method. But the curve can also be regarded by itself as a good criterion of the load of impurity in the water.

The weakness of the method lies in the establishment of an obligatory species standard.

The best results are obtained by longitudinal section investigations of larger rivers and streams which show long stretches of a uniform physiographic character. In smaller flowing waters, the physiographic conditions and consequently the living communities, change more and more markedly as the source is approached. It is difficult to select in these a suitable species standard.

3. Methods for Distinguishing Self-purification

Two further methods will now be mentioned which are concerned with the relationships between the reducers, consumers, and primary producers, and thus characterize the course of the self-purification.

The RPC system of Gabriel

Gabriel (1946) followed out the relationships of bacteria, ciliates, and chlorophyll-bearing plants to one another. The reducers R predominate in the heaviest contamination, then the consumers C are more numerous and with increasing self-purification the autotrophic producers P gain ground. Gabriel calculates a "biological index" from the quantitative relationships according to the formula:

$$I = \frac{2P}{R+C}.$$

The "Biological Index of Pollution" (BIP) *of Horasawa*

The Japanese worker Horasawa proceeds from conceptions similar to those of Gabriel but omits the use of bacteria. He works with chlorophyll-bearing organisms A and organisms without chlorophyll B. The relationship of these two to one another is characteristic of the biological situation during self-purification:

$$\text{BIP} = \frac{A}{A+B} \times 100.$$

Because this procedure was taken up by the World Health Organization (WHO) in its summary of the *International Standards for Drinking Water* (1958), at least a brief mention of it seems to be necessary here.

II. PHYSIOLOGICAL METHODS

A series of procedures for the biological estimation of water quality have been worked out which are based on the defined physiological characteristics of the test organisms. They cannot here be described in detail, but some of the procedures should be mentioned.

Oxygen consumption in 48 hours (BOD_2) and in 5 days (BOD_5):

The water sample is tested for its oxygen content, and then put into flasks for 48 hr or 5 days and kept in the dark at 20°C. After this time the final value of the oxygen is determined and compared with the initial value. The difference gives the oxygen consumption during this time. This value is proportional to the amount of putrescible substances in the water, so that conclusions can be drawn about the organic contaminations, especially if the procedure is combined with an estimation of the potassium permanganate consumption. In carrying out BOD_5 many methodological difficulties must be borne in mind, and these can be found in the technical books on water analysis.

Knöpp's "supplementary consumption"

The usual BOD procedure determines only the degree to which organic substances in the water sample are decomposed. Knöpp (1964) added to the water sample definite amounts of oxidizable substances and determined, so to speak, the oxidative self-purification potential on the basis of the supplementary consumption ZZ. Peptone ($ZZ_{peptone}$) was added as a type-substance for protein, and glucose ($ZZ_{glucose}$) for carbohydrate.

Determination of the sediment activity according to Caspers

10 ml of sediment are added to 1 l. of water saturated with oxygen and free from waste, and kept for 24 hr in the dark at 20°C. After this time the oxygen content of the water is determined. The deficit has been consumed by the sediment. The higher the content of oxidizable organic substances in the sediment, the greater is the consumption. Error $\pm 2{\cdot}45\%$. The procedure is very simply carried out.

The Biomass-Titre (BMT) method of Bringmann and Kühn

Sterile, membrane-filtered water samples are inoculated with known test organisms which can use the free nitrogenous compounds in the water and increase in biomass in proportion to the supply of these nutrients. Because only unicellular test organisms are used, the increase in growth after a given time can be expressed in terms of the increase in the number of cells, which is photometrically measured. A suspension of diatomaceous earth *(kieselguhr)* in water is used as a substance for calibration. The biomass is stated in milligrams of *kieselguhr*. To determine the degree of saprobic conditions bacteria of the genus *Escherichia* are used, which can take up only non-mineral nitrogen compounds (corresponding to the heterotrophic phase of the self-purification). To determine the trophic status the green alga *Scenedesmus quadricauda*, which exclusively utilizes inorganic nitrogen compounds, is used (autotrophic phase of the self-purification).

Test procedures for the determination of the toxicity of water

Industrial waste water, and also domestic waste water, often contain poisonous substances which are decomposed with difficulty or not at all. Numerous procedures have been developed to test the toxic effect of water and also the tolerance limits by means of test organisms. The *Escherichia coli*, the *Pseudomonas fluorescens*, the *Microregma heterostoma*, and the *Scenedesmus* tests of Bringmann and Kühn, the A-Z test of Knöpp (1961), and also the TTC-test of Bruckstееg and Thiele (11959) depend on injury to specific metabolic capacities of microorganisms. Higher organisms also, such as daphnias, have been used for toxicological tests (*Daphnia* test), by which the damage done to the animals after a certain time was determined. Zahner (1962) used fish as test objects for testing the toxicity of motor fuels and oils.

The publications of Bick (1963) and those of Bringmann, Kühn and Lüdemann (1962) summarize the methods.

APPENDIX I

METHODS FOR THE FIXATION AND PRESERVATION OF GROUPS OF FRESHWATER ORGANISMS

A. ANIMAL GROUPS

Bryozoa. These must be narcotized before they are fixed. To do this, put them in a small vessel in water and sprinkle some menthol crystals on the surface of the water. Cover with a glass plate. After a few hours the animals are extended and motionless. Fix them in Bouin or 70% alcohol or 10% formalin. Keep in alcohol or formalin.

Coelenterata. Let them extend in a little water and then add an equal volume of hot Bouin's fluid. Transfer after a few minutes into cold Bouin† and leave for 12 hr in this. Wash several times in clean water and preserve in 30% alcohol. Fixation in chromium-osmium-acetic acid and preservation in 2% formalin solution is equally good. To demonstrate the stinging cells the living animal is flooded with a few drops of a mixture of 1 part of 0·02% osmic acid and 4 parts of 5% acetic acid.

Coleoptera. The larvae can be killed and preserved in 70% alcohol. The imagines are killed in acetic-ether-soaked wool, and immediately preserved in a mixture of 65 parts of absolute alcohol, 30 parts of distilled water, and 5 parts of glacial acetic acid. Smaller imagines can also ne killed in this mixture.

Collembola. Fix in hot 96% alcohol. Preserve in 70% alcohol. Clear for microscopic investigation with creosote or 5% caustic soda.

Crustacea. All groups are killed with 4% formalin solution. Preserve in 96% alcohol with which 40 ml/l of glycerin has been mixed in order to prevent complete drying of the specimens. Zooplankton samples, which are to be counted, are better preserved in 4% formalin solutions or transferred to formalin before counting.

Diptera. Larger larvae are killed and fixed (Pennak) by flooding them with boiling water. Smaller larvae and the imagines are fixed and preserved in 70% alcohol; 4% formalin is also suitable.

Gastrotricha. The animals are put on to a microscope slide in water and inverted over a bottle containing 2% osmic acid; or fix in saturated mercuric chloride solution or 10% formalin or Bouin. Preserve in 70% alcohol or 5% formalin.

Gordidae. Fix in saturated mercuric chloride solution + 5–10% acetic acid (this is especially suitable for histological studies). Preserve in 90% alcohol. For collections also fix in 90% alcohol.

Hirudinea. Narcotize with carbon dioxide, chloroform, chloral hydrate, nicotine sulphate, etc., in aqueous solution, for 1–5 hr until the extended animals no longer react to pinching.

† 75 ml of saturated aqueous picric acid + 25 ml of 40% formaldehyde + 5 ml of glacial acetic acid.

Then fix in Schaudinn's solution† or in a mixture of 50% alcohol+2% formalin. Preserve in 80% alchohol or 5% formalin.

Hemiptera. Fix and preserve in 70% alcohol or pin out.

Hydracarina. Kill and preserve only in Koenike's mixture, consisting of glycerine, glacial acetic acid, and water in proportions of 10 : 3 : 6. Never kill or preserve in alcohol or formalin.

Lepidoptera. Fix and preserve the caterpillars and pupae in 70% alcohol. Kill the imagines in a potassium cyanide bottle and pin out.

Megaloptera and Neuroptera. Fix and preserve in 70% alcohol or pin out.

Mollusca, Snails. These are narcotized when they are extended. To do this some drops of a saturated solution of (cheap synthetic) menthol in 70% alcohol are poured on to the surface of the water: the menthol crystallizes out and forms a solid layer on the surface. Remove the shells. After $\frac{1}{2}$–2 days the outstretched animals no longer react to stimulation with a needle. Preserve in 70% alcohol, no formalin. (For histological study put the narcotized animals in Bouin.) When the animals are boiled in 5% caustic soda or caustic potash, the radula and jaws remain behind.

Mussels. These are killed by briefly treating them with boiling water (for small species: *Sphaeriidae*), or by boiling till the shell opens, and are preserved in 70% alcohol. No formalin.

Nematoda. After fixation by heat, it is best to preserve them in 5% formalin solution (Meyl). According to Meyl good results are also obtained with the following mixtures: (1) 1 ml of 40% formalin+1 ml of glacial acetic acid+8 ml of distilled water; (2) 100 ml of 96% alcohol+30 ml of 40% formalin +5 ml of glacial acetic acid+200 ml of distilled water.

Nemertina. Fix and preserve in 80% alcohol.

Odonata. Fix and preserve the larvae in 70% alcohol; the imagines are also preserved in alcohol or dried and pinned out.

Oligochaeta. Allow animals to crawl in a watch-glass with a little water: when they are fully extended flood them with hot Schaudinn's fluid.† Preserve in 80% alcohol containing a few drops of formalin. Naked species (*Nais*, *Stylaria*, *Dero*, and *Aeolosoma*) must be narcotized before fixation with a 2% solution of hydroxylamine hydrochloride.

Oribatidae. Fix and preserve in 96% alcohol, never in formalin.

Ostracoda. The whole sample is divided and half of it is immersed for a few seconds in boiling water (pieces of gauze or a small sieve). The shells of the animals thus treated gape open and can be much more easily detached for making a preparation. Preserve in formalin or alcohol.

Other Turbellaria. Before fixation examine living specimens and squash preparations, especially for Dalyellidae. Narcotize with 10% alcohol or 1% hydroxylamine. Fix with a solution of 2·5 g of potassium bichromate, 5 g of mercuric chloride, 1 g of sodium sulphate, 5 ml of formalin, and 100 ml of water. Remove the mercuric chloride with iodized alcohol and preserve in 70% alcohol. For histological examination fix in Bouin's mixture.

Plecoptera and Ephemeroptera. Fix and preserve the larvae and imagines in 70% alcohol: do not dry.

Porifera. Sponges are either dried or preserved in 70% alcohol. To prepare the spicules,

† Two parts of saturated aqueous solution of mercuric chloride and 1 part of 96% ethyl alcohol. After fixation wash thoroughly with iodized 70% alcohol.

small pieces of sponge are put into 2% nitric acid for 12 hr with frequent shaking. Let the spicules sediment out, pour off the solution, and wash them with water several times, and then in alcohol. Quick method: put some sponge tissue on a microscope slide with 2 drops of nitric acid and heat until dry. Then add mounting medium and put on a cover glass.

Protozoa. The treatment of the individual groups of Protozoa varies so much that it cannot be discussed in detail here.

Rotatoria (Rotifers). Species with which it is only a question of preserving the shell are fixed and preserved with 4% formalin. Species without a shell must be fixed extended. To do this a number of animals are put in a little water in a watch-glass and are flooded with about the same quantity of boiling water. Good results are also obtained by narcotization with a weak aqueous solution of cocaine. The following method is also very simple: some pieces of limestone are covered in a test-tube with some hydrochloric acid and the carbon dioxide that comes off is led into a second test-tube containing the sample of rotifers. The rotifers are narcotized in a few minutes and can be fixed in 4% formalin and preserved. Remane's method: add 4% formalin solution slowly drop by drop to the sample of rotifers. The animals are first narcotized and, when the concentration is increased, they are fixed. (Especially suitable for work in the field.)

Tardigrada. Fix and preserve in 85% alcohol or 5% formalin. For histological work fix in osmic acid (cf. Gastrotricha) or saturated aqueous mercuric chloride solution. Make holes in the dorsal skin so that the solution penetrates. The hard parts (mouth appendages, etc.) are placed on the microscope slide with 10% caustic soda.

Trichoptera. Fix and preserve the larvae and pupae in 70% alcohol and the imagines in 40–50% alcohol.

Triclads. Allow them to creep along in a little water and when they are extended cover them with Steinmann–Breslau's mercuric chloride – nitric acid.† After 1 min transfer to absolute alcohol and briefly remove the mercuric chloride with iodized alcohol. Preserve in 70% alcohol. For histological study use Bouin's mixture.

B. PLANT GROUPS

All groups, also net plankton, can be fixed with and preserved in 4% formalin. Algae can also be fixed and preserved with Pfeiffer's mixture: equal parts of wood vinegar, methyl alcohol, and water, but this has no advantage over formalin.

Water-bottle samples containing nannoplankton are fixed and preserved in Lugol's solution to which acetic acid or acetate has been added. The exact recipe is given on p. 58.

† One part of crude nitric acid, 1 part of concentrated mercuric chloride solution in salt water (5 g NaCl in 100 ml of water) and 1 part of distilled water.

APPENDIX II

SOME REMARKS ABOUT CULTURE METHODS

Bryozoa. The colonies cannot be kept well for a long time. Cultures are set up with pond water rich in algae, or the animals are fed with yeast. The temperature of the water should be 10–15°C, and if necessary aerate.

Cladocera. Woltereck (1908, 1909) fed Cladocera with algae or the genus *Chlorella*. Cultures of *Chlorella*, however, may be toxic because of the excretion of chlorellin in the plankton (Pratt *et al.*, 1940), as Ryther (1954) has also recently found for *Daphnia magna*. It is therefore best to use fresh cultures of *Chlorella* that are in active multiplication by division (see culture methods for algae, p. 164) or *Scenedesmus*, with which good culture results can be obtained. Daphnias also can be fed only on organic detritus and bacteria (Rodina, 1940, 1958; Rigler, 1958, 1961), but the use of bacteria in cultures is not to be recommended. Good results may, on the other hand, be obtained by feeding with yeast (von Dehn, 1950, 1955 with *Moina*). Daphnias especially feed excellently on yeast in open-air cultures (cement tanks).

Copepods. Lowndes (1928–9) obtained good culture results for Cyclopidae with hay infusions. Coker (1933) succeeded with suspensions of horse manure and chopped-up *Mougeotia* (a filamentous conjugate alga). Eckstein (1964) obtained good results with the calanoid copepods *Eudiaptomus gracilis* and *Mixodiaptomus laciniatus* by so arranging the cultures that he cultivated each batch of eggs laid in 50 ml of lake water (water in which they were found) and feeding them once a day with 4 drops of hay infusion and a mixed culture of *Scenedesmus*. The animals developed under these conditions to the adult stage and again laid eggs. The quality and quantity of the food, the quality of the water, and the temperature influence the development. It especially depends on scrupulously clean culture vessels which should never have had contact with chemicals, detergents, or soaps. Also it is no longer possible to use such vessels even with more careful cleaning. This is also true of Cladocera and for all cultures in general. Lepiney (1962) obtained good results with the American Cerophyl. One egg-bearing female was put into each Erlenmeyer (conical) flask, of 150 ml capacity containing 60 ml of Cerophyl solution, and the cultures were kept at 23°C for 8 months. They were fed with *Chlorella* and ciliates.

Hirudinea. It is easy to keep them, excepting *Piscicola*, because they only seldom need to be fed. The water must be cold and quite clear. Many substances, especially chlorine and copper, are injurious even in the smallest quantities; therefore do not use any tap water or copper piping. Dead animals, remains of food, and other impurities must be immediately removed. Glossiphonidae may be kept in smaller aquaria with a small amount of plant growth. For food, give live snails occasionally. Erpobdellidae must be kept in larger, covered aquaria with stones and mussel shells on which the cocoons can be deposited. Food: earthworms, insect larvae, scrapings of meat. Hirudidae are also kept in larger aquaria, but only in shallow water and with a layer of sand with moss and stones for drainage. Food: give one feed of vertebrate blood about every 6 months and between these feeds give earthworms, insect larvae, and frog spawn (from Pennak).

Hydrachnellae. Aquatic mites from standing waters may be easily kept for several months in the water in which they are found with a little mud and plant growth if sufficient food in the form of Cladocera, copepods, ostracods, and chironumid larvae is given to them. Soft-skinned species should not be kept with others because they are often attached and their tissues are sucked out. Species from springs and flowing waters and also those from ground water can be kept for a long time in shallow dishes, the bottoms of which are covered with mud. Temperature 10–15°C; feed as mentioned above. The cultivation of the species is difficult because their reproduction is very complex and most species pass through a parasitic stage on other animals (always, indeed, insects). For egg-laying, females are kept isolated in small aquaria, in which the eggs are laid on various substrates. (For this, see Sparing (1959), Bottger (1962), and identification books.) The limnohalacarid *Lobohalacarus weberi quadriporus*, which lives underground, has been cultured by Teschner (*Arch. Hydrobiol.* **59** (1963), 71 ff.) in Petri dishes 8·5 cm in diameter or in special microaquaria 1·7 cm in diameter and 1·5 cm deep. On the bottoms of the dishes was a sand layer (with a particle size of 0·14–0·32 mm or 0·14–1·02 mm) mixed with detritus and humus. Food: nematodes and pieces of oligochaetes. Cold shock should start the development of the eggs or even accelerate it.

Hydras. These are kept in small glass storage bottles filled with the water in which they were found. There should not be any aquatic plants or bottom material. Feed with Cladocera and copepods. For continuous culture give little food (proportion of hydras to daphnias to be 1 : 10). At temperatures of 20°C and with 3–4 daphnias per animal per day, vigorous reproduction occurs.

Insects. The culture and maintenance varies so much according to the species and groups of species and their biological requirements that it cannot be considered here. Specialized works on the individual groups give information (see References).

Naididae. Material from waters rich in plants is kept in aquaria or preserving jars together with the plants, especially *Elodea* and *Myriophyllum.* Naturally no aquatic animals, such as hydras, Turbellaria, ostracods, hydracarina, and young insect larvae (Odonata; *Lestes*), which feed on the Naididae, must be with them in the jars. Naididae very often multiply very rapidly in cultures of Protozoa.

Nematodes. The cultivation of the free-living aquatic species is difficult and, so far as I know, has not so far succeeded. For the culture of other species, see Meyl (1961).

Oligochaetes. Culture in shallow dishes with a weak current flowing through or water rich in oxygen dripping in. Never renew the water quickly. B. Wachs (1965) has published an interesting method for the quantitative sorting of oligochaetes out of muddy substrates.

Ostracoda. Many species may be easily kept for a long time in mud in shallow dishes if the conditions of the water in which they were found are reproduced. Animals from flowing waters and springs are kept in cold well-oxygenated water. Weygoldt (*Zool. Jb. Anat.* **78** (1960), 371) stated that the species *Cyprideis litoralis* (Brady), which prefers weakly saline water, produces eggs, when it is fed with stinging nettle powder, pulverized *Tubifex* or similar food. For further development of the eggs they are kept in small block-dishes. The addition of a harpacticid clears them of infection with bacteria and fungi and is also useful as a cleaner as it also is in the culture of the nauplii and later juvenile stages.

Rotatoria (rotifers). The culture and breeding of them is difficult. Most species must be fed with quite specific foods.

Epiphanes senta is fed with abundant infusoria or with *Euglena viridis* or with unicellular algae *(Chlorella, Raphidium)*, species of *Keratella* with *Chlorella*, and *Chromogaster* with *Ceratium hirundinella* or *Peridinium*. The bdelloid rotifer *Habrotocha constricta* is kept with an infusion of the moss *Polytrichum*. Many other bdelloid rotifers thrive in an infusion of the moss.

Hypnum schreberi. One should add to cultures of these rotifers a small quantity of powdered glass because the animals like to lay eggs between the granules of this. Rotifers creating eddies (with their trochal discs) may be kept for a long time in the water in which they were found, but this must be often renewed; the detritus in it serves as food. Bdelloid rotifers may even be observed for 2 months under a cover glass in this way (according to Donner).

Spongillidae. The animals are left for a short time in the aquarium. Only the water in which they are found must be used, and this must be renewed every 1–2 days.

Turbellaria. Triclads and rhabdocoeles from eutophic waters may be kept in shallow dishes filled with the water in which they were found. Rhabdocoeles are fed with *paramoecia*, small naids and daphnias; triclads with *Tubifex* or enchytraeid worms or with scrapings from beef liver. Feed only twice a week. After feeding remove the remains of the food and renew the water. Species from brooks and springs must be kept in shallow dishes at a low temperature with as much aeration as possible.

THE CULTURE OF ALGAE†

It is not all algae that can adequately supply their energy requirements by means of photosynthesis. Thus there is, for example, only a relatively scanty growth of *Euglenas*, if carbon dioxide is the only source of carbon supplied. Good growth of these species is only obtained if organic compounds such as acetate are given at the same time. Most colourless algae are dependent on the provision of an organic substrate. Even those forms, such as many strains of *Chlorella*, which reach a high rate of multiplication in purely mineral nutrient solutions, can thrive substantially better in the presence of glucose.

Some unicellular plankton algae are among the freshwater organisms which have been most studied and are most easily handled. Good growths in pure culture have also been obtained with numerous colonial and filamentous forms. However, the culture of the various groups of algae still always presents considerable difficulties.

Generally speaking, the culture conditions for algae in the laboratory are markedly different from those of their natural environments. This is especially true of the composition of the nutrient solution which must contain in a small volume nutrient substances in concentrations which supply the requirements of the alga growing in them during a long period of time. It is therefore necessary to remember that an alga taken from a culture differs in many characteristics from the species representatives living in the open.

Pringsheim (1946) has described how species of algae are isolated, and how absolutely pure cultures of them are made, and he also considers the specific requirements of the various groups of algae from the existing collections of pure cultures of algae, more than 2000 different strains being available. Some of the most extensive collections of this kind may here be named: the algal collections of the Institute of Plant Physiology of the University of Göttingen (Koch, 1964); the algal collection of the Botanical Institute of the University of Prague (Fott, 1964); and the collection of the Department of Botany, Indiana

† I am greatly indebted to Dr. C. J. Soeder, Falkau, for this summary and I incorporate it here word for word.

University, Bloomington (Indiana), U.S.A. (Starr, 1960). The references cited refer to the publication of the lists of species concerned.

Myers (1962), among others, has discussed the basis of the culture of algae for laboratory purposes. In order to obtain good results it is important to provide the correct salt concentration, pH value, intensity of illumination, and temperature. Algae can be cultivated either in nutrient solutions or on gelatinous nutrient media. The gelatinous nutrient media are especially suitable for the isolation of the organisms, for finding mutants, and for maintaining stock cultures.

The basis of most solid nutrient media is agar–agar obtained from red algae, but gelatin or silica gel are also used. Culture on such nutrient media also succeeds with numerous flagellates.

The primary requisite for the cultivation of pure cultures free from bacteria is to work under sterile conditions. If the nutrient solutions or media are sterilized in the autoclave, it must be remembered that this decomposes many organic compounds, so that the actual composition of the medium becomes substantially more complex than the composition stated in the formula for it. If the nutrient substrates contain organic compounds, sterilization by filtration is in many cases better than by autoclaving. Sterilization in the steam-chamber is also relatively mild. Under these conditions, in order to achieve a complete kill of fungal spores and bacteria, it is necessary to add soil extract and to repeat the steam sterilization about 24 hr later.

The literature contains abundant formulae for nutrient solutions, but the composition of these is only in individual instances based on experimental work. As an example of a medium with a relatively broad spectrum, the No. 10 nutrient solution of Chu (1942) may be mentioned:

Distilled water	1000 ml	Na_2CO_3	0·020 g
$Ca(NO_3)_2$	0·040 g	Na_2SiO_3	0·025 g
K_2HPO_4	0·010 g	$FeCl_3 \cdot 6H_2O$	0·008 g
$MgSO_4 \cdot 7H_2O$	0·025 g	Trace-element solution	1 ml

The trace-element solution contains in 1000 ml of distilled water:

$ZnSO_4 \cdot 7H_2O$	0·002 g	H_3BO_3	0·002 g
$MnSO_4 \cdot 4H_2O$	0·002 g	LiCl	0·001 g
$AlCl_3$	0·002 g	$CoCl_2 \cdot 6H_2O$	0·001 g

The concentration of nutrient solution of this kind is too weak to provide cultures that grow sufficiently quickly. For these, therefore, media of another composition are to be recommended, for example, the nutrient solution for *Chlorella* of Soeder and Ried (1963) which has the following composition:

Distilled water	1000 ml	$CaCl_2 \cdot 4H_2O$	0·001 g
KNO_3	1·00 g	$FeCl_3 \cdot 6H_2O$	0·001 g
$Na_2HPO_4 \cdot 12H_2O$	0·26 g	Trace-element solution modified after Hoagland (1938)	1 ml
KH_2PO_4	0·74 g		
$MgSO_4 \cdot 7H_2O$	0·05 g		

The trace-element solution contains in 1000 ml of distilled water:

$Al(SO_4)_3$	0·055 g	H_3BO_4	0·614 g
KI	0·028 g	$CuSO_4 \cdot 5H_2O$	0·055 g
KBr	0·028 g	$NiSO_4 \cdot 6H_2O$	0·055 g
LiCl	0·028 g	$Co(NO_3)_2 \cdot 6H_2O$	0·055 g
$MnCl_2 \cdot 4H_2O$	0·389 g	$ZnSO_4 \cdot 7H_2O$	0·055 g

This stock solution is diluted to 0·033 (1/30). The complete nutrient solution then contains 1 ml of the diluted trace-element solution.

In general a buffering of the culture medium is indicated. Good results are often achieved with phosphate buffer (see the formula given). Nevertheless, instances are known of high sensitivity to phosphate. Thus Rodhe (1948) found that even 5 g/l of phosphate was toxic to *Dinobryon divergens*. In such instances buffering of the nutrient solution with substances such as histidine or glycylglycine is recommended with the addition of phosphate in the form of organic eaters, i.e. perhaps glycerophosphate (Provasoli, 1960). Astonishing numbers of algae require vitamins (Droop, 1962). Culture of these succeeds only with the addition of vitamins, and almost always it is a question of thiamin biotin, vitamin B_{12}, or combinations of these.

The addition of chelators is either necessary or favourable to the culture of many algae. These organic compounds have the ability to take up heavy metal ions, which prevents their precipitation. Chelators suitable for algal cultures are, for example, ethylene diamine tetra-acetic acid (EDTA), amino acids such as histidine, or oligopeptides such as glycylglycine. The action of those substances depends, on the one hand, on the fact that they promote the uptake of trace-element cations, especially iron, or generally the algae to take up sufficient quantities of these. On the other hand, they neutralize the toxic effect of heavy metal ions by the formation of chelated complexes.

The favourable effect of soil extract (Pringsheim, 1946) on the growth of algae may be largely explained by its high content of chelators (especially humic acids) and by its vitamin content (especially of B_{12}). However, by the addition of soil extract to a nutrient solution, conditions result which cannot be accurately determined—conditions which are not desirable for accurate work. Another versatile method of providing vitamins, amino acids, and proteins, consists of the addition of yeast extract, which has an especially strong favourable effect on many algae which are not capable of photosynthesis. When amino acids are an obligatory requirement, the addition of peptone or casein is enough.

As an example of a nutrient solution which complies with many of the requirements just mentioned, a medium of Provasoli (1960) may be adduced, which was developed for the culture of *Synura*. A litre of the solution contains:

EDTA	50 mg	$ZnCl_2$	20·8 mg
$NaNO_3$	20 mg	$MnCl_2 \cdot 4H_2O$	7·2 mg
KH_2PO_4	14 mg	$(NH_4)_6Mo_7O_{24} \cdot 4H_2O$	0·13 mg
$MgSO_4 \cdot 7H_2O$	20 mg	$Co(NO_3)_2 \cdot 6H_2O$	0·13 mg
$CaCl_2 \cdot 2H_2O$	48 mg	NaH-glutamate	100 mg
KCl	10 mg	Na-acetate	40 mg
$FeCl_3 \cdot 6H_2O$	3·4 mg	pH 5·5	

High growth rates in cultures of photo-synthesizing algae were obtained only when sufficient carbon dioxide was available. To ensure this, blowing in mixtures of air and carbon dioxide may be more favourable than adding bicarbonate or carbonate to the nutrient solutions. An amount of carbon dioxide recommended is 1·0–1·5 vol. %. Gassing the culture has the additional substantial advantage that it constantly intermixes the algal suspension so that an ideal homogeneity of the conditions of growth can be achieved.

As concerns photosynthetic carbon dioxide assimilation, the algae may be cultivated directly in daylight, most favourably at a north window. But, because the intensity and quality of natural light do not last long and alter seasonally, illumination of cultures with artificial sources of light is generally preferable. For this purpose fluorescent tubes are usually used. Mixed light from daylight lamps and warm-tone lamps has also proved especially favourable. The intensity of illumination is arranged according to the growth capacity of the object. Species which grow slowly do better in weaker light, whilst rapid growing cultures (perhaps *Chlorella* and *Scenedesmus*) often just obtain optimal growth at 15 klux or more. Strong light can be injurious to sensitive algae or under unfavourable conditions of growth (low temperatures, insufficient supply of minerals, and the like) and this may lead to the death of the cultures.

Whilst many algae (e.g. coccoid green algae) can be kept directly in strong, continuous light, for many others a change from light to dark is more favourable and may even be absolutely necessary. This can be contrived in a simple manner by introducing a time switch in front of the source of light.

In algal cultures which are not kept permanently in a state of exponential growth by regular dilution, cell-multiplication and also substance production set in more or less rapidly and then they pass into the "stationary phase" (also the degeneration phase). This may cause fundamental changes in the chemical composition and in the physiological characteristics (von Witsch, 1950; Spoehr and Milner, 1949) and zooplankton fed with such algae may be injured (Ryther, 1954). In order to be able to reproduce always a definite physiological state and at the same time a definite composition of the algae, it is advisable to work under rigidly defined conditions. To do this it is best to use a light-thermostat (Lorenzen, 1959). The cultures should be otherwise diluted as regularly as possible to a constant density.

The technique of synchronous culture (Pirson, 1962; Lorenzen, 1964) is available to supply the most exacting requirements for reproducibility of the characteristics of plankton algae. Very promising, and especially suitable for experimental work on ecological problems, is homo-continuous culture with changes from light to dark by which a constant availability of light is also attained by automatic dilution (Senger, 1964).

Quantitative estimation of the growth capacity and production of substance by algae is possible only in liquid cultures. Reliable criteria for this are the dry weight, the protein content, or the number of cells. The last is usually determined with a haemocytometer (count chamber). But electronic cell-counters are also available (James, 1961). With suspensions of plankton algae, optical density can also be used as a measure of growth, but here it must be remembered that the particle size strongly influences the results of conventional photometric measurements. This undesirable factor is excluded by the opal glass method (Shibata, 1957).

When algae are cultivated to feed zooplankton, it should be remembered that the filtering freshwater Crustacea, for example, can feed only on very small nannoplankton. Small green algae seem to be less readily for this utilized than chrysomonads such as *Erkenia* and *Chromulina* (Nauwerck, 1963). Nevertheless, it is possible, in principle, to feed zooplankton with green algae (Ryther, 1954).

APPENDIX III

FIRMS WHICH MAKE HYDROBIOLOGICAL APPARATUS

FRANZ BERGMANN K. G., Bad Soden/Ts., Hauptstrasse 34, Germany.
HANS BÜCHI, Bern, Spital 18/Marktgasse 53, Switzerland.
ELECTROACUSTIC GmbH, Kiel, Westring 425/429, Germany.
KURT GOHLA, Kiel, Lornsenstrasse 43, Germany.
FRITZ HELLIGE & CO., GmbH, Freiburg i. Br., Heinrich-von-Stephan Strasse 4, Germany.
JOSEF HEPFINGER, München 26, Thierschstrasse 1, Germany; Miller's gauze for plankton nets.
BIOS HYDRO/APPARATEBAU GmbH, Kiel, Wismarerstrasse 14, Germany.
FRIEDRICH IHLENFELD, Hamburg-Bergedorf, Chrysanderstrasse 18, Germany; Miller's gauze for plankton nets.
MEMBRANFILTERGESELLSCHAFT, GmbH, Göttingen, Postfach 142, Germany.
A. OTT, Kempten/Allgäu, Mozartstrasse 18–20, Germany.
DR. BRUNO LANGE, Berlin-Zehlendorf, Hermannstrasse 14–18, Germany; electric light apparatus.
PHILIPP SCHENK, Wien XXI, Woltergasse 40, Austria.
SCHWEIZER SEIDENGAZE FABRIK A.G., Thal S.G., Switzerland.
DR. H. ZÜLLIG, APPARATEBAU, Rheineck S.G., Switzerland.

Other firms are given in: *Sources of Limnological and Oceanographic Apparatus and Supplies.* Amer. Soc. Limnol. Oceanogr. Spec. Publ. 1, (third edn.), April 1964.

REFERENCES

I. BOOKS AND COMPREHENSIVE PUBLICATIONS ON HYDROBIOLOGY AND LIMNOLOGY (op = out of print)

BERTSCH, K. (1947) *Der See als Lebensgemeinschaft*, O. Maier, Ravensburg, pp. 146.

BERTSCH, K. (1947) *Sumpf und Moor als Lebensgemeinschaft*, O. Maier, Ravensburg, pp. 125.

BREHM, V. (1930) *Einführung in die Limnologie*, Springer, Berlin, pp. 261 (op).

DUSSART, B. (1966) *Limnologie. L'Étude des eaux continentales*, Gauthier-Villars, Paris, pp. 677.

EDMONDSON, W. T. (1959) *Freshwater Biology*, Wiley, New York, 2nd edn., pp. 1248.

FREY, D. G. (1963) *Limnology in North America*, University of Wisconsin Press, Madison, pp. 734.

GESSNER, F. (1955) *Hydrobotanik*, Dt. Verl. Wiss., Berlin, vol. 1, pp. 517; vol. 2 (1959), pp. 701.

HARNISCH, O. (1929) Die Biologie der Moore, *Die Binnengewssäer* **7,** Schweizerbart, Stuttgart, pp. 146.

HARVEY, L. A. (1959) Ecological conditions in freshwater, *Sci. Progr.* **47,** 111–19.

HENTSCHEL, E. (1923) *Grundzüge der Hydrobiologie*, Fischer, Jena, pp. 221 (op).

HUTCHINSON, G. E. (1957) *A Treatise on Limnology*, vol. 1: *Geography, Physics and Chemistry*, Wiley, New York, pp. 1015.

HYNES, H. B. (1960) *The Biology of Polluted Waters*, Liverpool University Press, pp. 202.

ILLIES, J. (1961) Die Lebensgemeinschaft des Bergbaches, *Brehm-Bücherei*, **289,** Ziemsen, Wittenberg, pp. 106.

KELLER, R. (1961) *Gewässer und Wasserhaushalt des Festlandes*, Hauder & Spenersche Verlagsb., Berlin, pp. 520.

KUSNEZOW, S. J. (1959) *Die Rolle der Mikroorganismen im Stoffkreislauf der Seen*, translated from the Russian by A. Pochmann, Dt. Verl. Wiss., Berlin, pp. 301.

LANPERT, K. (1925) *Das Leben der Binnengewässer*, 3rd edn., Tauchnitz, Leipzig, pp. 892 (op).

LENZ, F. (1928) *Biologie der Süsswasserseen*, Springer, Berlin, pp. 221 (op).

LIEBMANN, H. (1954) Biologie und Chemie des ungestauten und gestauten Stromes, *Münchner Beiträge zur Abwasser-, Fischerei- und Flussbiologie*, **2,** Oldenbourg, Munich, pp. 315.

LIEBMANN, H. (1959) Methodik und Auswertung der biologischen Wassergüte-Kartierung, *Münchner Beiträge zur Abwasser-, Fischerei- und Flussbiology*, **6,** Oldenburg, Munich.

LIEBMANN, H. (1960) *Handbuch der Frischwasser- und Abwasser-Biologie*, Oldenbourg, Munich, vol. 2, pp. 1160; vol. 1, 2nd edn. (1962), pp. 588.

MACAN, T. T. (1961) Ecology of aquatic insects, *Ann. Rev. Ent.* **7,** 261–88.

MACAN, T. T. (1961) Factors that limit the range of freshwater animals, *Biol. Rev.* **36,** 151–98.

MACAN, T. T. (1963) *Freshwater Ecology*, Longmans, London, pp. 338.

MACAN, T. T. and WORTHINGTON, E. B. (1951) *Life in Lakes and Rivers*, Longmans, London, pp. 272.

MARGALEF, R. (1955) Los organismos indicatores en la limnología, *Biología de las aguas continentales*, **12,** pp. 300.

NAUMANN, E. (1931) Limnologische Terminologie, in ABDERHALDEN, *Handbuch der biologischen Arbeitsmethoden*, **IX,** 8, Urban & Schwarzenberg, Berlin and Vienna, pp. 776.

PAWLOWSKI, E. N. and SHADIN, W. J. (1950 to 1956) *Das Leben im Süsswasser der U.S.S.R.*, vol. 4, Moscow and Leningrad.

PENNAK, R. W. (1953) *Freshwater Invertebrates of the United States*, Ronald Press, New York, pp. 769.

PERES, J. M. and DEVEZE, L. (1963) *Océanographie biologique et biologie marine*. II, La vie pélagique, Press. Univ. Paris, pp. 511.

REID, K. G. (1961) *Ecology of Inland Waters and Estuaries*, Reinhold, New York, pp. 375.

REMANE, A. and SCHLIEPER, C. (1958) Die Biologie des Brackwassers, *Die Binnengewässer* **22,** Schweizerbart, Stuttgart, pp. 348.

RUTTNER, F. (1962) *Grundriss der Limnologie*, 3rd edn., de Gruyter, Berlin, pp. 332.

SCHWOERBEL, J. (1966) Ökologie der Süsswassertiere, Stehende Gewässer, *Fortschritte der Zoologie* **17** 389–427; **19,** 283–321 (1968).

SERNOW, S. A. (1958) *Allgemeine Hydrobiologie*, Dt. Verl. Wiss., Berlin, pp. 676.

STEUER, A. (1910) *Planktonkunde*, Teubner, Leipzig and Berlin, pp. 723 (op).

STEUER, A. (1911) *Leitfaden der Planktonkunde*, Teubner, Leipzig and Berlin, pp. 382 (op).

STRENZKE, K. (1957) Ökologie der Wassertiere, *Handbuch der Biologie*, **III**, Athenaion, 115–92, Constance.

THIENEMANN, A. (1925) Die Binnengewässer Mitteleuropas, *Die Binnengewässer* **1**, Schweizerbart, Stuttgart, pp. 255.

THIENEMANN, A. (1950) Verbreitungsgeschichte der Süsswassertierwelt Europas, *Die Binnengewässer* **18**, Schweizerbart, Stuttgart, pp. 809.

THIENEMANN, A. (1955) Die Binnengewässer in Natur und Kultur, *Verständliche Wissenschaft* **55**, Springer, Berlin, pp. 156.

THIENEMANN, A. (1956) Leben und Umwelt, *Rowohtz deutsche Enzyklopädie*, **22**, Hamburg, pp. 153.

WELCH, P. S. (1952) Limnology, McGraw-Hill, New York, pp. 538.

WESENBERG-LUND, C. (1939) *Biologie der Süsswassertiere (wirbellose Tiere)*, translated by Storch, O., Springer, Vienna, pp. 817.

WESENBERG-LUND, C. (1943) *Biologie der Süsswasserinsekten*, Springer, Berlin and Vienna, pp. 682.

II. PERIODICALS IN WHICH MAJOR HYDROBIOLOGICAL WORKS APPEAR

Acta Hydrobiologica, Polsk. Akad. Wiss., Cracow; commenced 1959.

Acta Limnologica, Carl Bloms, Lund; commenced 1948.

Annales Instituti Biologici (Tihany) Hungaricae Scientiarium, Akad. Nyomda, Budapest; commenced 1927.

Annales de Limnologie, Masson, Paris, commenced 1965.

Archiv für Fischereiwissenschaft, Vieweg, Braunschweig; commenced 1949.

Archiv für Hydrobiologie, with supplements: Tropische Binnengewässer, Falkau-Arbeiten, Elbe-Aestuar and Donauforschung, Schweizerbart, Stuttgart; commenced 1906.

Drottningholm Report, Carl Bloms, Lund; commenced 1933.

Ergebnisse der Limnologie, Schweizerbart, Stuttgart; commenced 1964.

Folia Limnologica Scandinavica, Einar Munksgaard, Copenhagen; commenced 1943.

Gewässer und Abwässer, Bagel, Dusseldorf; commenced 1953.

Hidrobiologia, Acad. Rep. Pop. Romine, Bucharest; commenced 1958.

Hydrobiologia, Junk, The Hague; commenced 1949.

Internationale Revue der gesamten Hydrobiologie, Akademie-Verlag, Berlin; commenced 1908.

Internationale Vereinigung für theoretische und angewandte Limnologie, Mitteilungen (published for Komitees für limnologische Methoden) Schweizerbart, Stuttgart; commenced 1953.

Jahrbuch "Vom Wasser", Verlag Chemie, Weinheim; commenced 1927.

Limnologica, Akademie-Verlag, Berlin; commenced 1962.

Limnology and Oceanography, Amer. Soc. Limnol. Oceanogr., Lawrence, Kansas; commenced 1956.

Meddelanden fran Lunds Universitets Limnogiska Institution, Carl Bloms, Lund; commenced 1948.

Memorie dell'Istituto Italiano die Idrobiologia "Dott. Marco De Marchi", Eigenverlag, Pallanza; commenced 1945.

Münchner Beiträge zur Abwasser-, Fischerei- und Flussbiologie, Oldenbourg, Munich; commenced 1927.

Polska Archivum Hydrobiologii, Polsk. Akad. Wiss., Warsaw; commenced ca. 1952.

Schweizerische Zeitschrift für Hydrologie, Birkhäuser, Basle; commenced 1920.

Scientific Papers from Institute of Chemical Technology, Stád. Pedag. Naklad., Prague; commenced 1957.

Verhandlungen der internationalen Vereinigung für theoretische und angewandte Limnologie, Schweizerbart, Stuttgart; commenced 1922.

Vie et Milieu, Hermann, Paris; commenced 1950.

Wasser und Abwasser, Winkler, Vienna; commenced 1958.

Zeitschrift für Fischerei und deren Hilfswissenschaften, Neumann, Berlin; commenced 1952.

III. COMPREHENSIVE PUBLICATIONS ON METHODOLOGY

ABDERHALDEN, E. (1923–1930), *Handbuch der biologischen Arbeitsmethoden*, **IX, 2, 5, 6, 8**, Methoden der Süsswasserbiologie und Meerwasserbiologie, Urban & Schwarzenberg, Berlin and Vienna. Abbreviated to ABDERHALDEN in the following references.

ASLO (1959) Sources of limnological and oceanographic apparatus ands upplies, *Limnol. Oceanogr.* **4**, 357–65.

AUERBACH, M. (1925) Fahrzeuge zur Untersuchung von Binnengewässern, ABDERHALDEN **IX, 2**, 463–98.

BIRGE, E. A. (1922) Second report on limnological apparatus, *Trans. Wisc. Acad. Sci. Arts Lettr.* **20,** 533–53.
ELSTER, H. J. (1961) Einige Bemerkungen zur Methodik der Limnologie, *Verh. int. Ver. Limnol.* **14,** 1115–18.
GROTE, A. (1955) Beitrag zur statistischen Überprüfung quantitativer Methoden in der Limnologie, *Gewässer und Abwässer* **5,** 31–76.
HALBFASS, W. (1923) *Grundzüge einer vergleichenden Seenkunde*, Bornträger, Berlin, pp. 354.
HALBFASS, W. (1922) Methoden der Seenforschung, ABDERHALDEN **X.**
HERBST, H. V. (1953) Biologische Arbeitsmethoden an Binnengewässern, *Gewässer und Abwässer* **3,** 7–30.
HOOWOOD, A. J. and OLIVE, J. R. (1961) Simple mechanical aids to limnological sampling, *Limnol. Oceanogr.* **6,** 87–88.
LUNDQUIST, G. (1925) Methoden zur Untersuchung der Entwicklungsgeschichte der Seen, ABDERHALDEN **IX, 2,** 427–62.
MORTON, F. (1931) Neue und verbesserte Instrumente zur Seenforschung, *Arch. Hydrobiol.* **22,** 484–6.
NAUMANN, E. (1925) Die Arbeitsmethoden der regionalen Limnologie, ABDERHALDEN **IX, 2,** 543–54.
NAUMANN, E. (1925) Methoden der experimentellen Aquarienkunde, ABDERHALDEN **IX, 2,** 621–52.
PESTA, O. (1931) Meine Hilfsmittel und Methoden bei der limnologischen Untersuchung von Hochgebirgsseen, *Arch. Hydrobiol.* **22,** 597–615.
PFANSTIEHL, A. (1962) An aid for small boat maneuvres, *Limnol. Oceanogr.* **7,** 431–2.
REINSCH, F. K. (1928) Limnologische Untersuchungen auf meiner Islandreise 1925. 2. Teil: Die Ausrüstung, insbesondere die neukonstruierten und verbesserten Instrumente, *Arch. Hydrobiol.* **19,** 383–420.
RIDLEY, J. E. and THURLEY, B. L. (1960) Limnological instruments and water supply, *Instrum. Eng.* **3,** 25–9.
SHADIN, W. I. (1960) *Methoden der Hydrobiologischen Forschung*, Moscow, pp. 191 (in Russian).
STEINER, G. (1919) Untersuchungsverfahren und Hilfsmittel zur Erforschung der Lebewelt der Gewässer, *Handb. mikr. Techn.* **7** and **8,** Franckh, Stuttgart, pp. 146.
WELCH, P. S. (1948) *Limnological Methods*, Blaikston, Philadelphia, pp. 381.

IV. PHYSICS AND CHEMISTRY

ALEKIN, O. K. (1962) *Grundlagen der Wasserchemie, Eine Einführung in die Chemie natürlichter Wasser*, VEB Dt. Verl. Grundstoffindustrie, Leipzig, pp. 260.
AMBÜHL, H. (1955) Die praktische Anwendung der elektrochemischen Sauerstoffbestimmung im Wasser, I, Grundlagen; Langzeitmessung, *Schweiz. Z. Hydrol.* **17,** 123–55; III, (1960) *Schweiz. Z. Hydrol.* **22,** 23–39.
AMERICAN PUBLIC HEALTH ASSOCIATION: *Standard Methods for the Examination of Water and Sewage*, 10th edn. (1955), translated from the 9th edn. by F. Sierp, (1951) Verlag Oldenbourg, Munich, pp. 522.
BAIER, C. (1945) Wasserschöpfer zur chemischen und bakteriologischen Probenentnahme aus tiefen und flachen Gewässern, *Arch. Hydrobiol.* **42,** 356–64.
BRESSLAU, E. (1925). Ein einfacher, für hydrobiologische, zoologische und botanische Zwecke geeigneter Apparat zur Messung der Wasserstoffionenkonzentration, *Arch. Hydrobiol.* **15,** 585–605.
CARPENTER, J. H. (1965) The accuracy of the Winkler method for dissolved oxygen analysis, *Limnol. Oceanogr.* **10,** 135–40.
CRAIG, R. E. and LAWRIE, R. G. (1962) An underwater light intensity meter, *Limnol. Oceanogr.* **7,** 259–61.
CRAIG, R. E. (1964) Discussion: radiation measurement in photobiology, choice of units, *Photochem. and Photobiol.* **3,** 189–94.
CZENSNY, R. (1926) *Die zu den wichtigsten chemischen Methoden der Wasseruntersuchung benötigten Gerätschaften und Chemikalien sowie ihre Anwendung auf der Reise und im Laboratorium*, Berlin.
DAM, L. van (1935) A method for determining the amount of oxygen dissolved in 1 ccm of water, *J. Exp. Biol.* **12.**
Deutsche Einheitsverfahren zur Wasseruntersuchung (1954) Verlag Chemie, Weinheim, pp. 180.
DIRMHIRN, I. (1964) *Das Strahlungsfeld im Lebensraum*, Akad. Verlagsges., Frankfurt/Main, pp. 426.
DORN, W. van (1957) Large-volume water sampler, *Trans. Amer. Geophys. Union* **37,** 682–4.
FINUCANE, J. H. and MAY, B. Z. (1961) Modified van Dorn water sampler, *Limnol. Oceanogr.* **6,** 85–87.
FITTKAU, E. J. (1957) Ein neuartiger Wasserschöpfer, *Z. Fisch. Hilfswiss.* N.F. **5,** 225–529.
FOX, H. M. and WINGFIELD, C. A. (1938) A portable apparatus for the determination of oxygen dissolved in a small volume of water, *J. Animal. Ecol.* **15.**
GÖRLICH, F. (1951) *Die Photozellen*, Akadem. Verl. Ges., Leipzig.
HAEGGBLOM, L. E. (1959) A thermistor thermometer which reads to the nearest 0,01° in the field, *Mitt. int. Ver. Limnol.* **10,** 1–24.

HASLAM, J., SQUIRREL, D. C. M., and BLACKWELL, J. G. (1960) The determination of calcium and magnesium in waters by automatic titration, *Analyst* **85,** 27–35.

HASSENTEUFEL, W., JAGITSCH, R. and KOCZY, F. F. (1963) Impregnation of glass surfaces against sorption of phosphate traces, *Limnol. Oceanogr.* **8,** 152–56.

HERON, J. (1962) Determination of phosphate in water after storage in polyethylene, *Limnol. Oceanogr.* **7,** 316–21.

HERON, J. and MACKERETH, F. J. H. (1955) The estimation of calcium and magnesium in natural waters, with particular reference to those of low alkalinity, *Mitt. int. Ver. Limnol.* **5,** 1–7.

HÖLL, K. (1967) *Untersuchung, Beurteilung, Aufbereitung von Wasser*, 3rd edn., de Gruyter, Berlin pp. 235.

JAAG, O., AMBÜHL, H., and ZIMMERMANN, P. (1956) Über die Entnahme von Wasserproben in fließenden Gewässern, *Schweiz. Z. Hydrol.* **18,** 156–60.

JORIS, L. S. (1964) A horizontal sampler for collection of samples near the bottom, *Limnol. Oceanogr.* **9,** 595–8.

KLEHN, H. and SONNTAG, D. (1963) Ein hydrographisches Extinktions- und Temperaturmessgerät, *Acta Hydrophys.* **8,** 23–45.

KLEINZELLER, A. (1965) *Manometrische Methoden und ihre Anwendung in der Biologie und Biochemie*, Gustav Fischer, Jena, pp. 620.

KLUT-OLSZEWSKI (1943) *Untersuchung des Wassers an Ort und Stelle*, Springer, Berlin, pp. 260.

KROGH, A. (1935) Syringe pipettes, *Ind. and Eng. Chem.* **27,** *Anal. Ed.* **7.**

KROGH, A. (1935) Precise determination of oxygen in water by syringe pipettes, ibid.

KROGH, A. (1941) *The Comparative Physiology of Respiratory Mechanisms*, Philadelphia, pp. 172.

KRUSE, H. (1949) *Wasser, Darstellung seiner chemischen, hygienischen, medizinischen und technischen Probleme*, Oppermann, Hannover, pp. 232.

LIEPOLT, R. (1960) Ein Profundalwasserschöpfer zur Erforschung der bodennahen Mikroschichtung stehender Gewässer, *Wasser und Abwasser*, 1960, 20–7.

LINDROTH, A. (1942) Mikromethoden für die hydrobiologische Feldarbeit, *Arch. Hydrobiol.* **38,** 436–45.

MACKERETH, F. J. H. (1957) Chemical analysis in ecology, illustrated from Lake District tarns and lakes: 1, Chemical analysis, *Proc. Linn. Soc. London*, Sess. **167,** 159–75.

MACKERETH, F. J. H. (1963) Some methods of water analysis for limnologists, *Freshwater Biol. Ass. Sci. Publ.* **21,** 1–70.

MACKERETH, F. J. H. (1964) An improved galvanic cell for determination of oxygen concentrations in fluids, *J. Sci. Instrum.* **41,** 38–41.

MAUCHA, R. (1932) Hydrochemische Methoden in der Limnologie, *Die Binnengewässer* **12,** 173.

MAUCHA, R. (1947) Hydrochemische Halbmikrofeldmethoden, *Arch. Hydrobiol.* **41,** 352–391.

MCLEOD, G. C., DOBBLIS, F., and YENTSCH, C. S. (1965) A graphic electrode system for measuring dissolved oxygen, *Limnol. Oceanogr.* **10,** 146–9.

MORTIMER, C. H. (1953) A review of temperature measurement in limnology, *Mitt. int. Ver. Limnol.* **1,** 1–25.

MORTIMER, C. H. (1956) The oxygen content of airsaturated fresh waters and aids in calculating percentage saturation, *Mitt. int. Ver. Limnol.* **6,** 1–20.

MORTIMER, C. H. and MOORE, W. H. (1953) The use of thermistors for the measurement of lake temperature, *Mitt. int. Ver. Limnol.* **2,** 1–42.

MÜLLER, G. (1964) *Methoden der Sedimentuntersuchung*, Schweizerbart, Stuttgart, pp. 303.

NOBEL, V. E. and AYERS, J. C. (1961) A portable photocell fluorometer for dilution measurements in natural waters, *Limnol. Oceanogr.* **6,** 457–61.

OHLE, W. (1932) Eine Selbstauslösevorrichtung zur Entnahme von bodennahen Wasserproben in Seen, *Arch. Hydrobiol.* **22,** 690–3.

PALLMANN, H., EICHELBERGER, E., and HASLER, A. (1940) Eine neue Methode der Temperaturmessung bei ökologischen und bodenkundlichen Untersuchungen, *Ber. schweiz. bot. Ges.* **50,** 337–62.

RICHARDSON, E. G. (1929) Two hot-wire viscometers, *J. Sci. Ind.* **6,** 11.

RIDLEY, J. E., ELLIOT, D. B. L., and OATEN, A. B. (1960) An automatic titrator as applied to Winkler's method for dissolved oxygen, *Analyst* **85,** 508–14.

RISCH, C. (1925) Ein brauchbarer Wasserschöpfer einfacher Konstruktion, *Arch. Hydrobiol.* **15,** 133–5.

RUTTNER, F. and HERRMANN, K. (1937) Über Temperaturmessungen mit einen neuen Modell des Lunzer Wasserschöpfers, *Arch. Hydrobiol.* **31,** 682–6.

SAUBERER, F. (1962) Empfehlungen für die Durchführung von Strahlungsmessungen an und in Gewässern, *Mitt. int. Ver. Limnol.* **11,** 77.

SAUBERER, F. and ECKEL, O. (1938) Zur Methodik der Strahlungsmessung unter Wasser, *Int. Rev. Hydrobiol.* **37,** 257–89.

SAUBERER, F. and RUTTNER, F. (1941) *Die Strahlungsverhältnisse der Binnengewässer*, Akad. Verlagsges., Leipzig, pp. 240.

SCHMASSMANN, H. (1956) Die Ermittlung der Sauerstoffsättigungskonzentration, *Schweiz. Z. Hydrobiol.* **18,** 144–55.

SCHMITZ, W. (1953) Die Anwendung optischer Messmethoden zur Erforschung der Binnengewässer, *Jahrb. v. Wasser* **20,** 127–36.

SCHMITZ, W. (1954) Grundlagen der Untersuchung der Temperaturverhältnisse in Fliessgewässern, *Ber. Limnol. Flussstat. Freudenthal* **6,** 29–50.

SCHMITZ, W. (1960) Lichtmessungen in Fliessgewässern des deutschen und österreichischen Donaugebietes, *Wetter und Leben* **12,** Sep., pp. 18.

SCHMITZ, W. and VOLKERT, E. (1959) Die Messung on Mitteltemperaturen auf reaktionskinetischer Grundlage mit dem Kreispolarimeter und ihre Anwendung in Klimatologie und Bioökologie, speziell in Forst- und Gewässerkunde, *Zeiss-Mitt.* **1,** 300–37.

STEEN, J. B. and IVERSON, O. (1965) Modernized Scholander respirometer for small aquatic animals, *Acta physiol. scand.* **63,** 171–4.

SUCHLAND, O. (1937) Methodisches zu Strahlungsmessungen unter Wasser, *Arch. Hydrobiol.* **31,** 201–8.

THIENEMANN, A. (1910) Eine einfache Form der Meyer'chen Schöpfflasche, *Arch. Hydrobiol.* **5,** 11–14.

TÖDT, F. (1958) *Elektrochemische* O_2*-Messungen*, de Gruyter, Berlin, pp. 212.

VOGLER, P. (1966) Analytik kondensierter Phosphate und organischer Phosphate bei limnologischen Untersuchungen, *Int. Rev. Hydrobiol.* **51,** 775–85.

WAGLER, E. (1932) Die chemische und physikalische Untersuchung der Gewässer für biologische Zwecke, ABDERHALDEN **IX, 2,** 2–72.

WESTLAKE, D. F. (1965) Some problems in the measurements of radiation under water: a review, *Photochem. and Photobiol.* **4,** 387–93.

V. PELAGIAL

AHLSTRØM, H. and THRAILKILL, J. R. (1962) Plankton volume loss with time of preservation, *Rapp. Proc.-verb. Cons. perm. int. Explor. Mer* **153,** 78.

APSTEIN, C. (1896) *Das Süßwasserplankton*, Lipsius & Tischer, Kiel and Leipzig, pp. 206.

ARON, W. (1962) Some aspects of sampling the macroplankton, *Rapp. Proc.-verb. Cons. perm. int. Explor. Mer* **153,** 29–38.

AURAND, K. and BEHRENS, H. (1966) Ein einfaches Gerät zur Sammlung und Entnahme von Sedimentproben aus Oberflächengewässern, *Arch. Hydrobiol.* **62,** 104–10.

BALLANTINE, D. (1953) Comparison of the different methods of estimating nannoplankton, *J. Mar. Biol. Assoc. U. K.* **32,** 129–47.

BANSE, K. (1962) Net zooplankton and total zooplankton, *Rapp. Proc.-verb. Cons. perm. int. Explor. Mer* **153,** 211–15.

BERGER, F. (1965) Bemerkungen zur graphischen Darstellung von Planktonvolksdichten, *Int. Rev. Hydrobiol.* **50,** 91–93.

BERND, D. and STEINECKE, H. (1955) Die Bedeutung der Plankton-Konzentration und deren quantitative Bestimmung in Oberflächengewässern, *Jahrb. v. Wasser* **22.**

BOLTOVSKOY, E. (1963) *Planktological Dictionary in Five Languages: English, Spanish, German, French, and Russian*, Buenos Aires, pp. 133.

BRAARUD, T. (1958) A: Methods for estimating standing crop in phytoplankton. A1: Counting methods for determination of the standing crop of phytoplankton, *Rapp. Proc.-verb. Cons. perm. int. Explor. Mer* **144,** 17–19.

BURCKHARD, G. (1910) Hypothesen und Beobachtungen über die Bedeutung der vertikalen Planktonwanderungen, *Int. Rev. Hydrobiol.* **3,** 156–72.

CLARK, W. J. and SIGLER, W. F. (1963) Method of concentrating phytoplankton samples using membrane filters, *Limnol. Oceanogr.* **8,** 127–9.

CLARKE, G. L. and BUMPUS, D. F. (1950) The plankton sampler—an instrument for quantitative plankton investigation, *Amer. Soc. Limnol. Oceanogr.*, Spec. Publ., **5,** 1–8.

CUSHING, C. E., Jr. (1961) A plankton subsampler, *Limnol. Oceanogr.* **6,** 489–90.

DAWSON, W. A. (1960) Home-made counting chambers for the inverted microscope, *Limnol. Oceanogr.* **5,** 235–6.

DJATSCHENKO, J. P. (1960) Preliminary results of the comparative investigations on the catch rate of the plankton gears, *Bull. Akad. Nauk SSSR*, **8/9,** 79–83 (in Russian).

DODSON, A. N. and THOMAS, W. H. (1964) Concentrating plankton in a gentle fashion, *Limnol. Oceanogr.* **9**, 455–6.

DUSSART, B. (1965) Les différentes catégories de plancton, *Hydrobiologia* **26**, 72–4.

ELSTER, H. J. (1956) Zur Methodik der Planktonforschung, *Publ. Staz. Zool. Napoli* **28**, 250–4.

ELSTER, H. J. (1958) Zum Problem der quantitativen Methoden in der Zooplanktonforschung, *Verh. int. Ver. Limnol.* **13**, 961–73.

ELSTER, H. J. and BRANDT, A. v. (1956) Unterwasser-Fernseh-Versuche im Königssee, *Arch. Hydrobiol., Suppl.*, **24**, 86–97.

FIELD, C. W. (1898) Use of the centrifuge for collecting plankton, *Science* **7**.

FLEMINGER, A. and CLUTTER, R. J. (1965) Avoidance of towed nets by zooplancton, *Limnol. Oceanogr.* **10**, 96–104.

FOXTON, P. (1963) An automatic opening-closing device for large plankton nets and midwater trawls, *J. Mar. biol. Ass. U. K.* **43**, 295–308.

GLENK, H. O. (1962) Pflanzliches Plankton. Methoden der wissenschaftlichen Planktonuntersuchung I–V, *Mikrokosmos* **51**, 178–82; 207–11; 268–71; 338–42; **52**, 203–7.

GLOVER, R. S. (1962) The continuous plankton recorder, *Rapp. Proc.-verb. Cons. perm. int. Explor. Mer* **153**, 59–65.

GUSEVA, K. A. (1959) About technics of phytoplankton studies, *Arb. Inst. Biol. Wod.* **2** (**5**), 44–51 (in Russian).

HANSEN, V. K. and ANDERSEN, K. P. (1962) Sampling the smaller zooplankton, *Rapp. Proc.-verb. Cons. perm. int. Explor. Mer* **153**, 39–47.

HEINRICH, K. (1934) Atmung und Assimilation im freien Wasser, *Int. Rev. Hydrobiol.* **30**, 387–410.

HENSEN, V. (1887) Über die Bestimmung des Planktons oder des im Meere treibenden Materials an Pflanzen und Tieren, 5. *Ber. Kommiss. wiss. Unters. Dt. Meere* **12–16**, 1–108.

HENSEN, V. (1895). *Ergebnisse der Planktonexpedition der Humboldtstiftung*, Lipsius & Tischer, Kiel and Leipzig.

HERBST, H. V. (1957) Der Fallschöpfer, ein Gerät zur quantitativen Zooplankton-Fang, *Arch. Hydrobiol.* **53**, 598–603.

HOPKINS, T. L. (1962) A zooplankton subsampler, *Limnol. Oceanogr.* **7**, 424–6.

HUNGER, H. (1957) Unterwasser-Fernsehen und seine Bedeutung für die Fischerei, *Allgem. Fischwirtschaftsztg.* **38**.

JACOBS, J. (1961) Cyclomorphosis in *Daphnia galeata mendotae* Birge, a case of environmentally controlled allometry, *Arch. Hydrobiol.* **58**, 7–71.

JACOBS, J. (1962) Light and turbulence as codeterminants of relative growth rates in cyclomorphic *Daphnia*, *Int. Rev. Hydrobiol.* **47**, 146–56.

JACOBS, W. (1935) Das Schweben der Wasserorganismen, *Ergebnisse der Biologie* **11**, 131–218.

JAVORNICKY, P. (1958) Die Revision einiger Methoden zum Festellen der Quantität des Phytoplanktons, *Sci. Pap. Inst. Chem. Technol. Prague, Fuel and Water* **2**, 283–367.

KNOTT, P. (1953) Modified whirling apparatus for the subsampling of plankton, *Austral. J. Mar. Freshw. Res.* **4**, 387–93.

KOLKWITZ, R. (1907) Entnahme- und Beobachtungsinstrumente für biologische Wasseruntersuchungen, *Mitt. Prüf.-Anst. Wasservers. Berlin* **9**, 111–14.

KOLKWITZ, R. (1911) Über das Kammerplankton des Süsswassers und der Meere, *Ber. dt. bot. Ges.* **29**, 386–412.

KOLKWITZ, R. (1911) Das Planktonsieb aus Metall und seine Anwendung, *Ber. dt. bot. Ges.* **29**.

KOLKWITZ, R. (1932) Die Planktonkammer unter Verwendung stärkerer Objektive, *Ber. dt. bot. Ges.* **50**, 374–5.

KRÄMER, A. (1897) Die Messung des Planktons mittels der Zentrifuge, *Verh. Ges. Naturf. Ärzte.* **68**.

KRIZENECKY, J. (1945) Untersuchungen zur Frage einer quantitativen Bestimmung des Teichplanktons mittels Zentrifugierens, *Arch. Hydrobiol.* **40**, 98–113.

KUTKUHN, J. H. (1958) Notes on the precision of numerical and volumetric plankton estimates from small samples concentrate, *Limnol. Oceanogr.* **3**, 69–83.

LAEVASTU, T. (1962) The adequacy of plankton sampling, *Rapp. Proc.-verb. Cons. perm. int. Explor. Mer* **153**, 66–73.

LOEFFLER, R. J. (1954) A new method of evaluating the distribution of planktonic algae in freshwater lakes, thesis, Univ. of Wisconsin, Madison.

LOHMANN, H. (1908) Untersuchung zur Feststellung des vollständigen Gehalts des Meeres an Plankton, *Wiss. Meeresuntersuch* **10**, 131–370.

LOHMANN, H. (1911) *Über das nannoplankton und die Zentrifugierung kleinster Wassermengen*, Leipzig.

LOVEGROVE, T. (1962) The effect of the various factors on dry weight values, *Rapp. Proc.-verb. Cons. perm. int. Explor. Mer* **163,** 86–91.

LUND, J. W. G. (1951) Sedimentation technique for counting algae and other organisms, *Hydrobiologia* **3,** 390–4.

LUND, J. W. G. (1959) A simple counting chamber for nannoplankton, *Limnol. Oceanogr.* **4,** 57–65.

LUND, J. W. G. (1962) Concerning a counting chamber described previously, *Limnol. Oceanogr.* **7,** 261–2.

LUND, J. W. G., KIPLING, C., and LE CREN, E. D. (1958) The inverted microscope method of estimating algal numbers and the statistical basis of estimation by counting, *Hydrobiologia* **11,** 143–70.

LUND, J. W. G. and TALLING, J. F. (1957) Botanical limnological methods with special reference to the algae, *Bot. Rev.* **23,** 489–583.

MCARTHUR, J. (1947) A new type of research microscope, *Trans. Roy. Trop. Med. Hyg.* **40,** 378.

MCARTHUR, J. (1947) Advances in the design of the inverted prismatic microscope, *J. Roy. Micr. Soc.*, Ser. 3, **65,** 8–16.

MONTI, R. (1911) Un noveau petit filet pour les pêches planktoniques de surface à toute vitesse, *Int. Rev. Hydrobiol.* **3,** 548–52.

MOORE, E. W. (1952) The precision of microscopic counts of plankton in water, *J. Amer. Wat. Wks. Assoc.* **44,** 208–16.

MOTADA, S. (1962) Plankton sampler for collecting uncontaminated materials from several different zones by a single vertical haul, *Rapp. Proc.-verb. Cons. perm. int. Explor. Mer* **153,** 55–8.

MOTADA, S. (1963) Devices of simple plankton apparatus, II, *Bull. Fac. Fish Hokkaido Univ.* **14,** 152–62.

NAUMANN, E. (1922) Über eine spezielle Anwendung der Zentrifugentechnik in der Planktonkunde, *Z. wiss. Mikros.* **39.**

NAUMANN, E. (1923) See und Teich (Plankton und Neuston), ABDERHALDEN **IX, 2,** 139–227.

NAUMANN, E. (1929) Grundlinien der experimentellen Planktonforschung, *Die Binnengewässer* **6,** pp. 100.

NAUWERCK, A. (1963) Die Beziehungen zwischen Zooplankton und im See Erken, *Symb. Bot. Upsal.* **XVII, 5,** 1–163.

OHLE, W. (1959) Blick in die Tiefe des Grossen Plöner Sees mit Fernseh- und Photokameras, *Natur und Volk* **87,** 177–88.

OSTWALD, W. (1902) Zur Theorie des Planktons, *Biol. Zbl.* **22,** 596–605, 609–38.

PALMER, C. M. and MALONEY, T. E. (1954) A new counting slide for nannoplankton, *Amer. Soc. Limnol. Oceanogr.*, Spec. Publ., **21,** 1–6.

PAQUETTE, R. G. and FROLANDER, H. F. (1957) Improvements in the Clarke–Bumpus plankton sampler, *J. Cons. int. Explor. Mer* **23,** 284–88.

PAQUETTE, R. G., SCOTT, E. L., and SUND, P. N. (1961) An enlarged Clarke–Bumpus plankton sampler, *Limnol. Oceanogr.* **6,** 230–3.

PASCHER, A. (1912) Versuche zur Methode des Zentrifugierens bei der Gewinnung des Planktons, *Int. Rev. Hydrobiol.* **5.**

PENNAK, R. W. (1962) Quantitative zooplankton sampling in litoral vegetation areas, *Limnol. Oceanogr.* **7,** 487–9.

RICKER, W. E. (1937) Statistical treatment of sampling process useful in the enumeration of plankton organisms, *Arch. Hydrobiol.* **31,** 68–84.

RINGELBERG, J. (1961) A physiological approach to an understanding of vertical migration, *Proc. Acad. Sci. Amsterdam* Ser. C, **64,** 489–500.

RINGELBERG, J. (1964) The positively phototactic reaction of *Daphnia magna* Straus, *Nederl. J. Sea Res.* **2,** 319–406.

RISCH, C. (1924) Ein Verfahren zur Zählung der Planktonorganismen, *Arch. Hydrobiol.* **15,** 416–21.

RODHE, W. (1941) Zur Verbesserung der quantitativen Phytoplanktonmethodik, *Zool. bidr. Uppsala* **20,** 465–77.

RUTTNER, F. (1909) Über die Anwendung von Filtration und Zentrifugierung bei den planktologischen Arbeiten an den Lunzer Seen, *Int. Rev. Hydrobiol.* **2,** 174–81.

RUTTNER, F. (1914) Die Verteilung des Planktons im Süßwasser, *Fortschr. naturw. Forsch.* **10;** ABDERHALDEN, **IX.**

RUTTNER, F. (1938) Limnologische Studien an einigen Seen der Ostalpen, *Arch. Hydrobiol.* **32,** 167–319.

RUTTNER, F. (1952) Planktonstudien der Deutschen Limnologischen Sundaexpedition, *Arch. Hydrobiol.*, Suppl., **10,** 1–274.

RYLOV, W. M. (1924) Zur Methodik der Untersuchungen des Kammerplanktons in sehr seichten Gewässern, Schr. *Süßwasser- u. Meereskde.*

RYLOV, W. M. (1927) Über die unmittelbare Verwendung der "KOLKWITZschen Planktonkammer" zur Entnahme des Planktons aus verschiedenen Tiefen, *Arch. Hydrobiol.* **18,** 60–64.

RYLOV, W. M. (1929) Anleitung zur Untersuchung des Limnoneustons, ABDERHALDEN **IX, 2,** 1385–418.

RYLOV, W. M. (1935) Das Zooplankton der Binnengewässer, *Die Binnengewässer* **15,** Schweizerbart, Stuttgart, pp. 212.

RYZEN, M. (1956) A microphotometric method of cell enumeration within the cerebral cortex of man, *J. Compar. Neurol.* **104.**

SCHÄRFE, J. (1952) Über Form und Größe von Wirkbereichen bei Fischlotungen, *Fischereiwelt* **2.**

SCHMIDT-RIES, H. (1936) Grundsätzliches zur Zentrifugenmethode, *Arch. Hydrobiol.* **29,** 553–616.

SCHRÖDER, R. (1959) Die Vertikalwanderungen des Crustaceenplanktons der Seen des südlichen Schwarzwaldes, *Arch. Hydrobiol.*, Suppl., **25,** 1–43.

SCHRÖDER, R. (1961) Untersuchungen über die Planktonverteilung mit Hilfe der Unterwasser-Fernsehanlage und des Echographen, *Arch. Hydrobiol.*, Suppl., **25,** 228–41.

SCHRÖDER, R. (1962) Vertikalverteilung des Zooplanktons und Thermokline, *Arch. Hydrobiol.*, Suppl., **25,** 401–10.

SCHRÖDER, R. (1962) Keine endogene Rhythmik bei den Vertikalwanderungen des Zooplanktons, *Arch. Hydrobiol.*, Suppl., **25,** 411–13.

SCHRÖDER, R. (1962) Vertikalverteilung des Zooplanktons in Abhängigkeit von den Strahlungsverhältnissen in Seen mit unterschiedlichen Eigenschaften, *Arch. Hydrobiol.*, Suppl., **25,** 414–29.

SCHRÖDER, R. and SCHRÖDER, H. (1964) Ein Beitrag zur Horizontalverteilung des Zoo- und Phytoplanktons in Querprofilen, *Mem. Ist. Ital. Idrobiol.* **17,** 81–101.

SCHRÖDER, R. and SCHRÖDER, H. (1964) On the use of the echo sounders in lake investigations, *Mem. Ist. Ital. Idrobiol.* **17,** 167–88.

SHIRAISHI, Y. (1960) On the application of echosounders to limnological research, *J. Japan. Limnol. Soc.* **21,** 245–55.

SIEBECK, O. (1960) Untersuchungen über die Vertikalwanderung planktischer Crustaceen unter Berücksichtigung der Strahlungsverhältnisse, *Int. Rev. Hydrobiol.* **45,** 381–454.

SIEBECK, O. (1964) Researches on the behaviour of planktonic crustaceans in the litoral, *Verh. int. Ver. Limnol.* **15,** 746–51.

SIEBECK, O. (1964) Experimente im Litoral zum Problem der "Uferflucht" planktischer Crustaceen, *Zool. Anz.*, Suppl., **27,** 388–96.

SOLI, G. (1964) A system for isolating phytoplancton organisms in unialgal and bacteria-free culture, *Limnol. Oceanogr.* **9,** 265–8.

STEEMANN NIELSEN, E. (1933) Über quantitative Untersuchungen von marinem Plankton mit Utermöhls umgekehrtem Mikroskop, *J. Cons. int. Explor. Mer* **8,** 201–10.

STEEMANN NIELSEN, E. and BRANDT, T. v. (1934) Quantitative Zentrifugenmethoden zur Planktonbestimmung, *Rapp. Proc.-verb. Cons. int. perm. Explor. Mer* **89,** 87–99.

ŠTEPÁNEK, M. (1961) Der automatische Planktonentnahmeapparat für Talsperren (HYDRA), *Verh. int. Ver. Limnol.* **14,** 955–7.

STRAŠKRABA, M. (1964) Preliminary results of a new method for the quantitative sorting of freshwater net plancton into main groups, *Limnol. Oceanogr.* **9,** 268–70.

SWANSON, G. A. (1965) Automatic plancton sampling system, *Limnol. Oceanogr.* **10,** 149–52.

THEILER, A. (1914) Ein neuer Wasser- und Planktonschöpfer nach Friedinger, *Int. Rev. Hydrobiol.*, Suppl. **6,** 1–4.

THOMAS, E. A. (1958) Das Plankton-Test-Lot, ein Gerät zum Studium des Verhaltens von Planktonorganismen im See, *Monatsbull. Schweiz. Ver. Gas- u. Wasserfachmännern* **38,** 1–6.

THOMASSON, K. (1963) Die Kugelkurven in der Planktologie, *Int. Rev. Hydrobiol.* **47,** 627–8.

TRANTER, D. J. and HERON, A. C. (1965) Filtration characteristics of Clarke-Bumpus samplers, *Aust. J. Mar. Freshwater Res.* **16,** 281–91.

UTERMÖHL, H. (1931) Über das umgekehrte Mikroskop, *Arch. Hydrobiol.* **22,** 643–5.

UTERMÖHL, H. (1932) Neue Wege in der quantitativen Erfassung des Planktons, *Verh. int. Ver. Limnol.* **5,** 567–95.

UTERMÖHL, H. (1936) Quantitative Methoden zur Untersuchung des Nannoplanktons, ABDERHALDEN **IX, 2/II,** 1879–937.

UTERMÖHL, H. (1958) Zur Vervollkommnung der quantitativen Phytoplankton-Methodik, *Mitt. int. Ver. Limnol.* **9,** 1–38.

VOLLENWEIDER, R. and WOLFF, H. (1949) Zur Methodik der Planktonstatistik, *Schweiz. Z. Hydrol.* **11,** 369–80.

WERFF, A. VAN DER (1955) A new method of concentrating and cleaning diatoms and other organisms, *Verh. int. Ver. Limnol.* **12,** 276–7.

WESENBERG-LUND, C. (1900) Über das Abhängigkeitsverhältnis zwischen dem Bau der Planktonorganismen und dem spezifischen Gewicht des Süsswassers, *Biol. Zbl.* **20,** 606–19, 644–50.

WIBORG, K. F. (1962) Estimations of number in the laboratory, *Rapp. Proc.-verb. Cons. perm. int. Explor. Mer.* **153,** 74–77.

WICKSTEAD, J. H. (1965) *An Introduction to the Study of Tropical Plankton,* Hutchinson, London.

WOOD, E. J. F. (1962) A method for phytoplankton study, *Limnol. Oceanogr.* **7,** 32–35.

WUNDER, W. (1935) Die Planktonröhre aus Cellhorn ein neues hydrobiologisches Gerät, *Arch. Hydrobiol.* **28,** 659–62.

YENTSCH, C. S., GRICE, G. D., and HART (1962) Some opening-closing devices for plankton nets operated by pressure, electrical and mechanical action, *Rapp. Proc.-verb. Cons. perm. int. Explor. Mer* **153,** 59–65.

ZACHARIAS, O. (1907) Der planktonseiher "Ethmophor", *Arch. Hydrobiol.* **2,** 320–4.

ZACHARIAS, O. (1907) *Das Süsswasserplankton,* Teubner, Leipzig.

VI. LITTORAL AND PROFUNDAL

ALM, G. (1924) Die quantitative Untersuchung der Bodenfauna und -flora in ihrer Bedeutung für die theoretische und angewandte Limnologie, *Verh. int. Ver. Limnol.* **2,** 1–12.

ALSTERBERG, G. (1922) Om prortagning med Ekman's bott-hämtare, *Skrift. Södr. Sver. Fiskeriför.* Lund.

ANDERSON, R. O. (1959) A modified flotation technique for sorting bottom sampler fauna, *Limnol. Oceanogr.* **4,** 223–5.

AUERBACH, M. (1953) Ein quantitativer Bodengreifer, *Beitr. naturkundl. Forsch. SW-Deutschl.* **12,** 17–23.

AURAND, K. and BEHRENS, H. (1966) Ein einfaches Gerät zur Sammlung und Entnahme von Sedimentproben aus Oberflächengewässern, *Arch. Hydrobiol.* **62,** 104–10.

BARTHELMESS, D. (1963) Über die horizontale Wanderung der Tiefenfauna, *Z. Fisch. Hilfswiss.* N. F. **11,** 183–7.

BAYES, J. C. and ANSELL, A. D. (1964) A simple automatic bottom water sampler for shore use, *Limnol. Oceanogr.* **9,** 600–1.

BERG, K. (1937) Contribution to the biology of *Corethra* Meigen (*Chaoborus* Lichenstein), *Kgl. Danske Selskh. Biol. Medd.* **13,** 1–101.

BIRKETT, L. (1957) Flotation technique for sorting grab samples, *J. Cons. int. Explor. Mer* **22,** 289–92.

BÜLOW, T. v. (1951) Die Seerosenzone als Lebensraum, Diss. Univ. Kiel.

BURBANECK, W. D. and ALLEN, J. M. (1947) A simple method of collecting small sessile freshwater forms, *Turtox News* **25,** 241–3.

CAVENESS, F. E. and JENSEN, H. J. (1955) Modification of the centrifugal-flotation technique for the isolation and concentration of nematodes and their eggs from soil and plant tissue, *Proc. Helm. Soc. Wash.* **22,** 87–89.

CHOLNOKY, B. v. (1927) Untersuchungen über die Ökologie der Epiphyten, *Arch. Hydrobiol.* **18,** 661–705.

COOKE, W. B. (1958) Continuous sampling of trickling filter populations, I, Procedures, *Sewage Industr. Wastes* **30,** 21–27.

DILLON, W. P. (1964) Flotation technique for separating fecal-pellets and small marine organisms from sand, *Limnol. Oceanogr.* **9,** 601–2.

DITTMAR, H. (1952) Anwendungsmöglichkeiten des zerhackten Gleichstroms für ökologische Arbeiten, *Arch. Hydrobiol.* **45,** 217–23.

EHRENBERG, H. (1957) Die Steinfauna der Brandungsufer ostholsteinischer Seen, *Arch. Hydrobiol.* **53,** 87–159.

EINSLE, U. (1964) Larvalentwicklung von Cyclopiden und Photoperiodik, *Die Naturwissenschaften* **51,** 345.

EKMAN, S. (1911) Neue Apparate zur qualitativen und quantitativen Untersuchung der Bodenfauna der Binnenseen, *Int. Rev. Hydrobiol.* **3,** 553–61.

ELGMORK, K. (1962) A bottom sampler for soft mud, *Hydrobiologia,* **20,** 167–72.

ELSTER, H. J. (1933) Eine Schlittendredge, *Int. Rev. Hydrobiol.* **29,** 290–2.

ELSTER, H. J. (1955) Limnologische Untersuchungen im Hypolimnion verschiedener Seetypen, *Mem. Ist. Ital. Idrobiol.,* Suppl., **8,** 83–119.

FORSBERG, C. (1959) Quantitative sampling of sub-aquatic vegetation, *Oikos* **10,** 233–40.

FRANKLIN, W. R. and ANDERSON, D. V. (1961) A bottom sediment sampler, *Limnol. Oceanogr.* **6,** 233–4.

FREIDENFELT, T. (1924) Neue Versuche zur Methodik der quantitativen Untersuchung der Binnenseen. Einige kritische Bemerkungen, *Arch. Hydrobiol.* **14,** 572–7.

FREMLING, C. R. (1961) Screened pail for sifting bottom-fauna samples, *Limnol. Oceanogr.* **6,** 96.

FREY, D. G. (1964) Remains of animals in quaternary lake and bog sediments and their interpretation, *Ergebn. Limnol.* **2,** 1–114.

FROLANDER, H. F. and PRATT, J. (1962) A bottom skinner, *Limnol. Oceanogr.* **7**, 104–6.
GAMS, H. (1925) Die Höhere Wasservegetation, ABDERHALDEN **IX, 2,** 713–50.
GARNETT, P. A. and HUNT, R. H. (1965) Two techniques for sampling freshwater habitats, *Hydrobiologia* **26,** 114–20.
GESSNER, F. (1952) Der Druck in seiner Bedeutung für das Wachstum submerser Wasserpflanzen, *Planta* **40,** 391–7.
GESSNER-LIERSCH, H. (1950) Zur Ökologie des Phragmitesgeleges, *Abh. Fischerei Lief.* **3,** 525–604.
GOLUBIČ, S. (1963) Hydrostatischer Druck, Licht und submerse Vegetation im Vrana-See, *Int. Rev. Hydrobiol.* **48,** 1–7.
GRZENDA, A. R. and BREHMER, M. L. (1960) A quantitative method for the collection and measurement of stream periphyton, *Limnol. Oceanogr.* **5,** 190–4.
GÜNTHER, B. (1963) Ein neuer Bodengreifer, *Z. Fisch. Hilfswiss.* N. F. **11,** 635–9.
HELL, W. (1960) Zur Methodik der quantitativen Gewinnung und Gewichtsbestimmung der Makrostein- und Schlammfauna, *Wasser und Abwasser*, 1960, 28–34.
HENTSCHEL, E. (1916) Biologische Untersuchungen über den tierischen und pflanzlichen Bewuchs im Hamburger Hafen, *Mitt. Zool. Mus. Hamburg* **33,** 1–172.
HENTSCHEL, E. (1923/4) Abwasseruntersuchungen mit biologischen Methoden im Hamburger Elbgebiet, *Techn. Gemeindeblatt* **26,** 113–15.
IVLEV, V. S. (1929) Zur Frage über den Bewuchs in den Teichen, *Arb. d. Hydrobiol. Stat. am See Glubokoje* **6,** 70–103 (in Russian).
IVLEV, V. S. (1933) Ein Versuch zur experimentellen Erforschung der Ökologie der Wasserbiocönosen, *Arch. Hydrobiol.* **25,** 177–91.
JONASSON, P. M. (1954) An improved funnel trap for capturing emerging aquatic insects, with some preliminary results, *Oikos* **5,** 170–89.
JONASSON, P. M. (1955) The efficiency of sieving techniques for sampling freshwater bottom fauna, *Oikos* **6,** 183–208.
JONASSON, P. M. (1958) The mesh factor in sieving techniques, *Verh. int. Ver. Limnol.* **13,** 860–7.
JUSE, A. (1966) Diatomeen in Seesedimenten, *Erg. d. Limnol.* **4,** 1–32.
KAJAK, Z. (1963) Analysis of quantitative benthic methods, *Ekol. Polsk.*, Ser. A, **11,** 1–56.
KANN, E. (1941) Cyanophyceenkrusten aus einem Teich bei Abisko (Schwedisch-Lappland), *Arch. Hydrobiol.* **37,** 495–503.
KANN, E. (1941) Krustensteine in Seen, *Arch. Hydrobiol.* **37,** 504–32.
KANN, E. (1941) Ökologische Untersuchungen an Litoralalgen ostholsteinischer Seen, *Arch. Hydrobiol.* **37,** 177–269.
KNUDSEN, M. (1927) A bottom sampler for hard bottoms, *Medd. Komm. Havunders. S. Fisk.* **8.**
KORDE, N. W. (1966) Algenreste in Seesedimenten, *Erg. d. Limnol.* **3,** 1–38.
KRASOVSKA, K. and MIKULSKI, S. (1960) Studies on animal aggregations associated with immersed and pleustonic vegetations in lake Druzuo, *Ecol. Polsk.*, Ser. A, **8,** 353–5.
LANG, K. (1930) Ein neuer Typ des quantitativen Bodenschöpfers, *Arch. Hydrobiol.* **21,** 145–50.
LELLACK, J. (1961) Zur Benthosproduktion und ihrer Dynamik in drei böhmischen Teichen, *Verh. int. Ver. Limnol.* **14,** 213–19.
LELLACK, J. (1961) Die Bedeutung der Chironomiden in unseren Teichen, *Z. Vesmir* **40,** 376–8.
LENHARD, G. (1966) A new elutriation sieve technique for the determination of texture and structure of bottom deposits of natural waters, *Arch. Hydrobiol.* **62,** 82–94.
LENZ, F. (1931) Untersuchung über die Vertikalverteilung der Bodenfauna im Tiefensediment von Seen. Ein neuer Bodengreifer mit Zerteilungsvorrichtung, *Verh. int. Ver. Limnol.* **5,** 232–61.
LENZ, F. (1932) Zur Methodik der quantitativen Bodenfauna-Untersuchung. Der Stockhalter, ein neues Hilfsgerät zum Bodengreifer, *Arch. Hydrobiol.* **23,** 375–80.
LIEBMANN, H. (1950) Zur Biologie der Methanbakterien, *Gesundheitsing.* **71.**
LÖFFLER, H. (1961) Vorschlag zu einem automatischen Schlämmverfahren, *Int. Rev. Hydrobiol.* **46,** 288–91.
LUNDBECK, J. (1926) Die Bodentierwelt norddeutscher Seen, *Arch. Hydrobiol.*, Suppl., **7,** 1–473.
LUNDBECK, J. (1954) Zur Kenntnis der Lebensvorgänge in sauren Binnenseen, *Arch. Hydrobiol.*, Suppl., **20,** 18–117.
MACAN, T. T. (1958) Methods of sampling the bottom fauna in stony streams, *Mitt. int. Ver. Limnol.* **8,** 1–21.
MESCHKAT, A. (1934) Der Bewuchs in den Röhrichten des Plattensees, *Arch. Hydrobiol.* **27,** 437–517.
MESCHKAT, A. (1934) Methoden der Bewuchsuntersuchungen an Schilfstengeln, *Arb. Ung. Biol. Forschungsinst. Tihany* **7,** 154–62.

Meuche, A. (1939) Die Fauna im Algenbewuchs. Nach Untersuchungen im Litoral ostholsteinischer Seen, *Arch. Hydrobiol.* **34**, 349–520.

Müller-Liebenau, I. (1956) Die Besiedlung der Potamogeton-Zone ostholsteinischer Seen, *Arch. Hydrobiol.* **52**, 470–606.

Mundie, J. H. (1956) Emergence traps for aquatic insects, *Mitt. int. Ver. Limnol.* **7**, 1–13.

Mundie, J. H. (1959) Diurnal activity of the larger invertebrates at the surface of Lac la ronge, Saskatchewan, *Canad. J. Zool.* **37**, 945–56.

Murray, J. W. (1962) A new bottom-water sampler for ecologists, *J. Mar. Biol. Assoc. U.K.* **42**, 499–501.

Neel, J. K. (1948) A limnological investigation of the psammon in Douglas Lake with a special reference to shoal and shoreline dynamics, *Trans. Amer. Micr. Soc.* **67**, 1–53.

Newcombe, C. L. (1949) Attachment materials in relation to water productivity, *Trans. Amer. Micr. Soc.* **68**, 355–61.

Newcombe, C. L. (1950) A quantitative study of attachment materials in Sodon Lake, Michigan, *Ecology* **31**, 204–15.

Northcote, T. G. (1964) Use of a high frequency echo sounder to record distribution and migration of *Chaoborus* larvae, *Limnol. Oceanogr.* **9**, 87–91.

Pennak, R. W. (1940) Ecology of the microscopic metazoa inhabiting the sandy beaches of some Wisconsin Lakes, *Ecol. Monogr.* **10**, 537–615.

Pennak, R. W. (1950) Comparative ecology of the interstitial fauna of fresh-water and marine beaches, *Ann. Biol.* **27**, 449–80.

Perfilew, B. W. (1927) Zur Methodik der Erforschung von Schlammablagerungen, Abderhalden **IX.**

Perfilew, B. W. (1929) Ein neuer Apparat für tiefe Schlammbohrungen vom Boote auf grösseren Tiefen (Kolbenbohrer), *Int. Hydrol. Kongr. Sevilla 1929.*

Pieczyńska, E. (1960) Types settlement by water mites *(Hydracarina)* of the littoral zone of Lake Wilkus, *Ekol. Polska*, Ser. B, **6**, 339–46.

Pieczyńska, E. (1964) Investigations on colonisation of new substrates by nematodes *(Nematoda)* and some other periphyton organisms, *Ekol. Polska*, Ser. A, **XII**, 185–214.

Ponyi, J. E. (1960) Über im interstitialen Wasser der sandigen und steinigen Ufer des Balaton lebenden Krebse (Crustacea), *Annal. Biol. Tihany* **27**, 85–92.

Ponyi, J. E. (1962) Zoologische Untersuchung der Röhrichte des Balaton, I, Krebse (Crustacea), *Annal. Biol. Tihany* **29**, 129–63.

Rawson, D. S. (1947) An automatic-closing Ekman-dredge and other equipment for use in extremely deep water, *Limnol. Amer. Soc. Limnol. Oceanogr.*, Spec. Publ., **18.**

Rawson, D. S. (1953) The bottom fauna of Great Slave Lake, *J. Fish. Res. Board Canada* **10**, 486–520.

Reineck, H. E. (1962) Der Kastengreifer, *Natur und Museum* **93**, 102–8.

Reissinger, A. (1930) Methoden zur Untersuchung von Seeschlammschichten, ihrer Mächtigkeit und ihrer Zusammensetzung, *Ber. Naturw. Ges. Bayreuth* **3.**

Rieth, A. (1960) Ein vereinfachter funktionssicher Bodengreifer, *Mikrokosmos* **49**, 252–6.

Rieth, A. (1961) Der Pflanzengreifer, eine zum Sammeln der Unterwasservegetation geeignete Abwandlung des Schlammschöpfers, *Mikrokosmos* **50**, 61–63.

Roll, H. (1939) Zur Terminologie des Periphytons, *Arch. Hydrobiol.* **35**, 59–69.

Ruttner-Kolisko, A. (1953) Psammonstudien I, Das Psammon des Torneträsk in Schwedisch-Lappland, *Sitz. Ber. Österr. Akad. Wiss.* I, **162**, 129–161.

Ruttner-Kolisko, A. (1954) Psammonstudien II, Das Psammon des Erken in Mittelschweden, *Sitz. Ber. Österr. Akad. Wiss.* I, **163**, 301–24.

Ruttner-Kolisko, A. (1962) Porenraum und kapillare Wasserströmung im Limnopsammal, ein Beispiel für die Bedeutung verlangsamter Strömung, *Schweiz. Z. Hydrol.* **24**, 444–58.

Sander, G. (1957) Beitrag zur Genauigkeit der Bodengreifer-Methodik, *Z. Fisch. Hilfswiss.* N.F. **6**, 251.

Schäperclaus, W. (1939) Eine nützliche kleine Verbesserung am Bodengreifer von Ekman-Birge, *Arch. Hydrobiol.* **35**, 169–70.

Schlee, D. (1965) Ein automatisches Fangkarussell zur Freiland-Untersuchung der tageszeitlichen Schlüpfrhythmik bei Wasserinsekten, *Arch. Hydrobiol.* **61**, 215–27.

Schönborn, W. (1962) Zur Ökologie der Testaceen im oligotrophen See, dargestellt am Beispiel des Grossen Strechlinsees, *Limnologica* **1**, 111–82.

Schräder, T. (1932) Über die Möglichkeit einer quantitativen Untersuchung der Boden- und Ufertierwelt fliessender Gewässer, *Z. Fischerei* **30**, 105–27.

Sládeček, V. and Sládečková, A. (1963) Relationship between wet weight and dry weight of the periphyton, *Limnol. Oceanogr.* **8**, 309–11.

SLÁDEČKOVÁ, A. (1960) Application of the glass slide method to the periphyton study in the Slapy Reservoir, *Sci. Pap. Inst. Chem. Technol. Prague, Fuel and Water* **4,** 403–34.

SLÁDEČKOVÁ, A. (1960) Limnological study of the reservoir Sedliče near Zeliv, XI, Periphyton stratification during the first year-long period, ibid. **4,** 143–261.

SLÁDEČKOVÁ, A. (1962) Limnological investigation methods for the periphyton (Aufwuchs) community, *Bot. Rev.* **28,** 286–350.

SMITH, A. J. (1959) Description of the Mackereth portable core sampler, *J. Sediment. Petrol.* **29,** 246–50.

STEINMANN, P. and SURBECK, G. (1922) Zum Problem der biologischen Abwasseranalyse, *Arch. Hydrobiol.* **13,** 404–14.

SZLAUER, L. (1963) Diurnal migrations of minute invertebrates inhabiting the zone of submerged hydrophytes in a lake, *Schweiz. Hydrol.* **25,** 56–64.

THIENEMANN, A. (1918) Untersuchungen über Beziehungen zwischen dem Sauerstoffgehalt des Wassers und der Zusammensetzung der Fauna in den norddeutschen Seen, 1, Mitteilung, *Arch. Hydrobiol.* **12,** 1–65.

THIENEMANN, A. (1923) See und Teich (Ufer), ABDERHALDEN **IX, 2,** 97–138.

THOMASSON, H. (1925) Methoden zur Untersuchung der Mikrophyten der Limnischen Litoral- und Profundalzone, ABDERHALDEN **IX, 2,** 681–712.

TONOLLI, V. (1962) Nuovi strumenti per la racolta e la separazione dei popolamenti benthonici, *Publ. Staz. Zoo. Napoli Suppl.* **32,** 20–29.

WASMUND, E. (1932) Entwicklung und Verbesserung der Entnahmeapparatur für Bodenproben unter Wasser, *Arch. Hydrobiol.* **22,** 646–62.

WESENBERG-LUND, C. (1908) Die litoralen Tiergesellschaften unserer grösseren Seen I, *Int. Rev. Hydrobiol.* **1,** 574–609.

WILLER, A. (1923) Der Aufwuchs der Unterwasserpflanzen, *Verh. int. Ver. Limnol.* **1,** 37–57.

WÜLKER, W. (1961) Lebenszyklus und Vertikalverteilung der Chironomide (Dipt.) *Sergentia coracina* Zett im Titisee, *Verh. int. Ver. Limnol.* **14,** 962–7.

YOUNT, J. L. (1956) Factors that control species numbers in Silver Springs, Florida, *Limnol. Oceanogr.* **1,** 286–95.

YUNKER, C. J. (1959) An improved method for storage and shipment of small invertebrate specimens, *Turtox News* **37,** 294–5.

ZAHNER, R. (1964) Beziehungen zwischen dem Auftreten von Tubificiden und der Zufuhr organischer Stoffe im Bodensee, *Int. Rev. Hydrobiol.* **49,** 417–54.

ZÜLLIG, H. (1953) Ein neues Lot zur Untersuchung der obersten Schlammschichten, zur Messung des Sedimentabsatzes und zur Erfassung bodennaher Wasserschichten, *Schweiz. Z. Hydrol.* **15,** 275–84.

ZÜLLIG, H. (1956) Sedimente als Ausdruck des Zustandes eines Gewässers, *Schweiz. Z. Hydrol.* **18,** 5–143.

ZÜLLIG, H. (1959) Eine neue Schöpfflasche mit automatischer Bodenauslösevorrichtung, *Schweiz. Z. Hydrol.* **21,** 109–11.

VII. RESEARCH ON RUNNING WATER

ALBRECHT, M.-L. (1952) Die Plane und andere Flämingbäche, *Z. Fischerei,* N. F. **1,** 389–476.

ALBRECHT, M.-L. (1959) Die quantitative Untersuchung der Bodenfauna fließender Gewässer, *Z. Fischerei,* N. F. **8,** 481–550.

ALBRECHT, M.-L. (1961) Ein Vergleich quantitativer Methoden zur Untersuchung der Makrofauna fließender Gewässer, *Verh. int. Ver. Limnol.* **14,** 486–90.

ALLANSON, B. R. and KERRICH, J. E. (1961) A statistical method for estimating the number of animals found in field samples drawn from polluted rivers, *Verh. int. Ver. Limnol.* **14,** 491–94.

ALLEN, K. R. (1940) Studies on the biology of the early stages of the salmon *(Salmo salar)*, 1, Growth in the river Eden, *J. Anim. Ecol.* **9,** 1–23.

ALM, G. (1922) Über die Prinzipien der quantitativen Bodenfaunistik und ihre Bedeutung für die Fischerei, *Verh. int. Ver. Limnol.* **1,** 168–80.

AMBÜHL, H. (1959) Die Bedeutung der Strömung als ökologischer Faktor, *Schweiz. Z. Hydrol.* **21,** 133–264.

BACKHAUS, D. (1967) Ökologische und experimentelle Untersuchungen an den Aufwuchsalgen der Donauquellflüsse Breg und Brigach und der obersten Donau bis zur Versickerung bei Immendingen, *Arch. Hydrobiol.*, Suppl., **30,** 364–99.

BADCOCK, R. M. (1953) Comparative studies in the populations of streams, *Rep. Inst. Freshwater Res. Drottningholm* **35,** 38–50.

BAGNOLD, R. A. (1951) Measurements of very low velocities of water-flow, *Nature* **167.**

BEAK, T. W. (1938) Methods of making and sorting collections for an ecological study of a stream, *Avon Biolog. Res.*, Ann. Rep. 1936–1937, **5**, 42–46.

BEHNING, A. (1928) Das Leben der Wolga, *Die Binnengewässer* **5**, (Schweizerbart), Stuttgart, pp. 162.

BELING, D. (1929) La faune aquatique des fleuves méridionaux de l'Ukraine en rapport avec la question de son origine, *Verh. int. Ver. Limnol.* **4**, 213–39.

BERG, K. (1948) Biological studies on the River Susaa, *Fol. Limnol. Scand.* **4**, 1–318.

BEYER, H. (1932) Die Tierwelt der Quellen und Bäche des Baumbergegebietes, *Abh. Westf. Prov. Mus. Naturkde.* **3**, 1–185.

BORNHAUSER, K. (1912) Die Tierwelt der Quellen und Bäche in der Umgebung Basels, *Int. Rev. Hydrobiol. Biol.*, Suppl., **5**, 1–90.

BUTCHER, R. W. (1933) Studies on the ecology of rivers, I. On the distribution of macrophytic vegetation in the rivers of Britain, *J. Ecol.* **21**, 58–91.

BUTCHER, R. W. (1938) Algae on the river, *Avon Biol. Res.*, Ann. Rep. 1936–7, **5**, 47–52.

BUTCHER, R. W. (1949) Problems of distribution of sessile algae in running water, *Verh. int. Ver. Limnol.* **10**, 98–103.

CHUTTER, F. M. and NOBLE, R. G. (1966) The reliability of a method of sampling stream invertebrates, *Arch. Hydrobiol.* **62**, 95–103.

CUSHING, C. E. (1964) An apparatus for sampling drifting organisms in streams, *J. Wildlife Management* **28**, 592–4.

DITTMAR, H. (1955) Die quantitative Analyse des Fließwasser-Benthos, *Arch. Hydrobiol.*, Suppl., **22**, 295–300.

DITTMAR, H. (1955) Ein Sauerlandbach. Untersuchungen an einem Wiesen-Mittelgebirgsbach, *Arch. Hydrobiol.* **50**, 307–544.

DODDS, G. S. and HISAW, F. L. (1925) Ecological studies on aquatic insects, IV. Altitudinal range and zonation of mayflies, stoneflies, and caddisflies, *Ecology* **6**, 380–90.

DORRIS, T. C. (1961) A plankton sampler for deep river waters, *Limnol. Oceanogr.* **6**, 366–7.

DOUGLAS, B. (1958) The ecology of the attached diatoms and other algae in a small stony stream, *J. Ecol.* **46**, 295–322.

EDINGTON, J. M. and MOLYNEAUX, L. (1960) Portable water velocity meter, *J. Sci. Instrum.* **37**, 455–7.

ELSTER, H.-J. (1962) Seetypen, Fließgewässertypen und Saprobiensystem, *Int. Rev. Hydrobiol.* **47**, 211–18.

ENGELHARDT, W. (1951) Faunistisch-ökologische Untersuchungen über Wasserinsekten an den südliche Zuflüssen des Ammersees, *Mitt. Münch. Entomol. Ges.* **41**, 1–135.

GEJSKES, D. C. (1935) Faunistisch-ökologische Untersuchungen am Röserenbach bei Liestal im Baseler Tafeljura, *T. Entomol.* **78**, 251–382.

GERSBACHER, W. M. (1937) The development of stream bottom communities in Illinois, *Ecology* **18**, 359–90.

GESSNER, F. (1950) Die ökologische Bedeutung der Strömungsgeschwindigkeiten fliessender Gewässer und ihre Messung auf kleinstem Raum, *Arch. Hydrobiol.* **43**, 149–65.

GLEDHILL, T. (1960) The *Ephemeroptera*, *Plecoptera* and *Trichoptera* caught by emergence traps in two streams during 1958, *Hydrobiologia* **15**, 179–88.

HARROD, J. J. (1962) The distribution of invertebrates on submerged aquatic plants in a chalk stream, with special reference to the Simuliidae, thesis, Univ. Southampton.

HENTSCHEL, E. (1923) Die Untersuchung von Strömen, ABDERHALDEN **IX**, **2**, 87–96.

HUBAULT, E. (1927) Contribution à l'étude des invertébrés torrenticoles, *Bull. Biol. France et Belg.*, Suppl., **9**, 1–388.

HYNES, H. B. (1961) The invertebrate fauna of a Welsh mountain stream, *Arch. Hydrobiol.* **57**, 344–88.

IDE, F. P. (1940) Quantitative determination of the insect fauna of rapid water, *Univ. Toronto biol.*, Ser. **47**, 5–20.

ILLIES, J. (1952) Die Mölle. Faunistisch-ökologische Untersuchungen an einem Forellenbach im Lipper Bergland, *Arch. Hydrobiol.* **46**, 424–612.

ILLIES, J. (1953) Die Besiedlung der Fulda (insbes. das Benthos der Salmonidenregion) nach dem jetzigen Stand der Untersuchung, *Ber. Limnol. Flussstat. Freundenthal* **5**, 1–28.

ILLIES, J. (1955) Der biologische Aspekt der limnologischen Fliesswassertypisierung, *Arch. Hydrobiol.*, Suppl., **22**, 337–46.

ILLIES, J. (1958) Die Barbenregion mitteleuropäischer Fliessgewässer, *Verh. int. Ver. Limnol.* **13**, 834–44.

ILLIES, J. (1961) Versuch einer allgemeinen biozönotischen Gliederung der Fliessgewässer, *Int. Rev. Hydrobiol.* **46**, 205–13.

ILLIES, J. and BOTOSANEANU, L. (1963) Problèmes et méthodes de la classification et de la zonation écologique des eaux courantes, considérés surtout du point de vue faunistique, *Mitt. int. Ver. Limnol.* **12**, 1–57.

JONASSON, P. M. (1948) Quantitative studies of the bottom fauna, in BERG, Biological studies on the river Susaa, *Fol. Limnol. Scand.* **4**, 203–84.

KAMLER, E. and RIEDEL, W. (1960) A method for quantitative study of the bottom fauna of Tatra streams, *Polsk. Arch. Hydrobiol.* **8**, 95–105.

KNÖPP, H. (1960) Untersuchungen über das Sauerstoff-Produktions-Potential von Flussplankton, *Schweiz. Z. Hydrol.* **22**, 152–66.

KOLKWITZ, R. and MARSSON, M. (1908) Ökologie der pflanzlichen Saprobien, *Ber. Deutsch. Bot. Ges.* **26**, 505–19.

KOLKWITZ, R. and MARSSON, M. (1909) Ökologie der tierischen Saprobien, *Int. Rev. Hydrobiol.* **2**, 126–52.

KOTHÉ, P. (1961) Hydrobiologie der Oberelbe. Natürliche, industrielle und wasserwirtschaftliche Faktoren in ihrer Auswirkung auf das Benthos des Stromgebietes oberhalb Hamburgs, *Arch. Hydrobiol.*, Suppl., **24**, 221–343.

LAMMERS, W. T. (1962) Density gradient separation of plankton and clay from river-water, *Limnol. Oceanogr.* **7**, 224–9.

LIEBMANN, H. (1954) Biologie der Donau und des Mains, *Münch. Beitr. Abw. Fisch. Flussbiol.* **2**, 111–209.

LYAKHOV, S. M. and ZHIDKOV, L. F. (1953) A bottom sampler—an apparatus for studying benthic organisms carried down by river current, *Zool. Zh.* **32**, 1020–4 (in Russian).

MACAN, T. T. (1958) Methods of sampling the bottom fauna in stony streams, *Mitt. int. Ver. Limnol.* **8**, 1–21.

MARGALEF, R. (1949) A new limnological method for the investigation of thin-layered epilithic communities, *Hydrobiologia* **1**, 215–16.

MARGALEF, R. (1949) Une nouvelle méthode limnologique pour l'étude de périphyton, *Verh. int. Ver. Limnol.* **10**, 284–5.

MARGALEF, R. (1960) Ideas for a synthetic approach to the ecology of running water. *Int. Rev. Hydrobiol.* **45**, 133–53.

MOON, H. P. (1935) Methods and apparatus suitable for an investigation of the litoral region of oligotrophic lakes, *Int. Rev. Hydrobiol.* **32**, 319–33.

MÜLLER, K. (1954) Beitrag zur Methodik der Untersuchung fliessender Gewässer, *Arch. Hydrobiol.* **43**, 567–70.

MÜLLER, K. (1965) An automatic stream drift sampler, *Limnol. Oceanogr.* **10**, 483–5.

MÜLLER, K. (1966) Die Tagesperiodik von Fliesswasserorganismen, *Z. Morph. Ökol. Tiere* **56**, 93–142.

MUNDIE, J. H. (1956) Emergence traps for aquatic insects, *Mitt. int. Ver. Limnol.* **7**, 1–13.

MUNDIE, J. H. (1964) A sampler for catching emerging insects and drifting materials in streams, *Limnol. Oceanogr.* **9**, 456–9.

NEEDHAM, P. R. (1934) Quantitative studies of stream bottom foods, *Trans. Amer. Fish. Soc.* **64**, 238–47.

NEEDHAM, P. R. and USINGER, R. L. (1956) Variability in the macrofauna of a single-riffle in Prosser Creek, California, as indicated by the Surber sampler, *Hilgardia* **24**, 383–409.

NEEDHAM, R. P. (1927) Biological survey of the Oswego river system, Suppl. to *17th Ann. Rep. N.Y. State Cons. Dept.*

NEIL, R. M. (1937–1938) The food and feeding of the brown trout (*Salmo trutta L.*) in relation to the organic environment. *Trans. Roy. Soc. Edinb.* **59**, II, 481–520.

NIETZKE, G. (1937) Die Kossau. Hydrobiologisch-faunistische Untersuchungen an schleswigholsteinischen Fliessgewässern, *Arch. Hydrobiol.* **32**, 1–74.

ODUM, H. T. (1957) Trophic Structure and Productivity of Silver Springs, Floride, *Ecol. Monogr.* **27**, 55–112.

PERCIVAL, E. and WHITEHEAD, H. (1929) A quantitative study of the fauna of some types of stream-bed, *J. Ecol.* **17**, 282–314.

PLESKOT, G. (1951) Wassertemperatur und Leben im Bach, *Wetter u. Leben* **3**, 129–43.

PLESKOT, G. (1953) Beiträge zur Limnologie der Wienerwaldbäche, *Wetter u. Leben*, Sonderh., **2**, 1–216.

PLESKOT, G. (1958) Die Periodizität einiger Ephemeropteren der Schwechat, *Wasser und Abwasser* 1958, 1–32.

REDEKE, H. C. (1923) Rapport omterent het woorkomen en den groi van jonge zalmpjes in zuidlimburgsche Beken, *Verh. rapp. rijksinst. visscherijonderzoek*, **I** (2), 183–220.

ROLL, H. (1938) Die Pflanzengesellschaften ostholsteinischer Fließgewässer, *Arch. Hydrobiol.* **34**, 159–305.

RUTTNER-KOLISKO, A. (1961) Biotop und Biozönose des Sandufers einiger österreichischer Flüsse, *Verh. int. Ver. Limnol.* **14**, 362–8.

SHELFORD, V. E. and EDDY, S. (1929) Methoden zur Untersuchung von Flußlebensgemeinschaften, ABDERHALDEN **IX, 2**, 1525–49.

SCHERER, E. (1965) Zur Methodik experimenteller Fließwasser-Ökologie, *Arch. Hydrobiol.* **61**, 242–8.

SCHMITZ, W. (1957) Die Bergbach-Zoozönosen und ihre Abgrenzung, dargestellt am Beispiel der oberen Fulda, *Arch. Hydrobiol.* **53**, 465–98.

SCHRÄDER, T. (1932) Über die Möglichkeit einer quantitativen Untersuchung der Boden- und Ufertierwelt fließender Gewässer, *Z. Fischerei* **30**, 105–27.

SCHWOERBEL, J. (1955) Ökologische Studien an torrentikolen Wassermilben *(Hydrachnellae, Acari)*. Ein Beitrag zur Ökologie unserer Schwarzwaldbäche, *Arch. Hydrobiol.*, Suppl., **22**, 530–7.

SCHWOERBEL, J. (1959) Die biologische Gliederung des Rheinstromes, *Gas- und Wasserfach* **100**, Sep. 1–6.

SCHWOERBEL, J. (1959) Ökologische und tiergeographische Untersuchungen über die Milben *(Acari Hydrachnellae)* der Quellen und Bäche des südlichen Schwarzwaldes und seiner Randgebiete, *Arch. Hydrobiol.*, Suppl., **24**, 385–546.

SCHWOERBEL, J. (1961) Über die Lebensbedingungen und die Besiedlung des hyporheischen Lebensraumes, *Arch. Hydrobiol.*, Suppl., **25**, 182–214.

SCHWOERBEL, J. (1961) Subterrane Wassermilben (Acari, Hydrachnellae, Porohalacaridae und Stygothrombiidae), ihre Ökologie und Bedeutung für die Abgrenzung eines aquatischen Lebensraumes zwischen Oberfläche und Grundwasser, *Arch. Hydrobiol.*, Suppl., **25**, 242–306.

SCHWOERBEL, J. (1964) Die Bedeutung des Hyporheals für die benthische Lebensgemeinschaft der Fließgewässer, *Verh. int. Ver. Limnol.* **15**, 215–26.

SCHWOERBEL, J. (1964) Die Wassermilben (Hydrachnellae und Limnohalacaridae) als Indikatoren einer biocönotischen Gliederung von Breg und Brigach sowie der obersten Donau, *Arch. Hydrobiol.*, Suppl., **27**, 386–417.

SLACK, K. V. (1955) A study of the factors affecting stream productivity by the comparative method, *Invest. Indiana Lakes and Streams* **4**, 3–47.

SLANINA, K. (1958) Die Verarmung von Fließwasserbiocoenosen durch Flotationsabgänge, *Wasser und Abwasser* 1958, 3–23.

SOMMERMANN, K. M., SALLER, R. I. and ESSELBAUGH, C. O. (1955) Biology of Alaskan black flies (Simuliidae, Diptera), *Ecol. Monogr.* **25**, 345–85.

SPENCER, E. A., TUDHOPE, J. S., and MORRIS, J. I. N. (1960) Flow measurements by the salt-dilution method, *J. Instt. Wat. Engrs.* **14**, 28–36.

SPRULES, W. M. (1947) An ecological investigation of stream insects in Algonquin Park, Ontario, *Univ. Toronto Stud.*, Biol. Ser., **56**, 1–81.

STEINMANN, P. (1907) Die Tierwelt der Gebirgsbäche, eine faunistisch-biologische Studie, *Ann. Biol. Lacustre* **2**, 30–164.

TANAKA, H. (1960) On the daily change of the drifting of benthic animals in streams, especially on the types of daily change observed in taxonomic groups of insects, *Bull. Freshwater Fish. Res. Lab. Tokyo* **9**, 13–26.

THIENEMANN, A. (1912) Der Bergbach des Sauerlandes. Faunistisch-biologische Untersuchungen, *Int. Rev. Hydrobiol.*, Suppl., **4**, 1–125.

THIENEMANN, A. (1926) Hydrobiologische Untersuchungen an den kalten Quellen und Bächen der Halbinsel Jasmund auf Rügen, *Arch. Hydrobiol.* **17**, 221–336.

TINDALL, D. R. and MINCKLEY, W. L. (1964) An integrated application of three kinds of sampling techniques to stream limnology, *Limnol. Oceanogr.* **9**, 270–2.

TÜMPLING, W. v. (1960) Probleme, Methoden und Ergebnisse biologischer Güteuntersuchungen an Vorflutern, dargestellt am Beispiel der Werra, *Int. Rev. Hydrobiol.* **45**, 513–34.

USINGER, R. L. and NEEDHAM, P. R. (1956) A drag-type riffle-bottom sampler, *The progressive Fish-culturist* **18**, 42–44.

WATERS, T. F. (1962) A method to estimate the production rate of a stream bottom invertebrate, *Trans. Amer. Fish. Soc.* **91**, 243–50.

WATERS, T. F. (1962) Diurnal periodicity in the drift of stream invertebrates, *Ecology* **43**, 316–20.

WHITFORD, L. A. and SCHUMACHER, G. J. (1963) Communities of algae in North Carolina streams and their seasonal relations, *Hydrobiologia*, **22**, 133–96.

WHITFORD, L. A., DILLARD, G. E., and SCHUMACHER, G. J. (1964) An artificial stream apparatus for the study of lotic organisms, *Limnol. Oceanogr.* **9**, 598–600.

WHITLEY, L. S. (1962) New sampler for use in shallow streams, *Limnol. Oceanogr.* **7**, 265–6.

WILDING, J. L. (1940) A new square-foot aquatic sampler, *Amer. Soc. Limnol. Oceanogr.*, Spec. Publ., **4**, 1–4.

ZAHNER, R. (1959) Die Bindung der Calopteryxarten an das strömende Wasser, *Int. Rev. Hydrobiol.* **44**, 51–130.

ZIMMERMANN, P. (1961) Der Einfluss der Strömung auf die Zusammensetzung der Lebensgemeinschaften im Experiment, *Schweiz. Z. Hydrol.* **23**, 1–81.

ZSCHOKKE, F. (1900) *Die Tierwelt der Gebirgsbäche,* Chur.

VIII. SUBTERRANEAN WATER

ANGELIER, E. (1953) Recherches écologiques et biogéographiques sur la faune des sables submergés, *Arch. Zool. expér. et gén.* **90**, 37–162.

ANGELIER, E. (1953) Le Peuplement des sables submergés d'eau douce, *Ann. Biol.* **29**, 467–86.

CHAPPUIS, P. A. (1922) Die Fauna der unterirdischen Gewässer in der Umgebung von Basel, *Arch. Hydrobiol.* **14**, 11–88.

CHAPPUIS, P. A. (1927) Die Tierwelt der unterirdischen Gewässer, *Die Binnengewässer* **3**, pp. 175.

CHAPPUIS, P. A. (1930) Methodik der Erforschung der subterranen Fauna, ABDERHALDEN **IX**, **2**.

CHAPPUIS, P. A. (1942) Eine neue Methode zur Untersuchung der Grundwasserfauna, *Act. Sci. math. nat. Univ. Francisco-Josephina Kolozsvar* **6**, 1–7.

DELAMARE-DEBOUTTEVILLE, C. (1960) *Biologie des eaux souterraines littorales et continentales*, Herrmann, Paris, pp. 740.

DICHTL, G. (1959) Die Grundwasserfauna im Salzburger Becken und im anschliessenden Alpenvorland, *Arch. Hydrobiol.* **55**, 281–370.

HAINE, E. (1946) *Die Fauna des Grundwassers von Bonn mit besonderer Berücksichtigung der Crustaceen*, Melle i. Hann., pp. 144.

HAMANN, O. (1896) *Europäische Höhlenfauna*, Costenoble, Jena, pp. 296.

HUSMANN, S. (1956) Untersuchungen über die Grundwasserfauna zwischen Harz und Weser, *Arch. Hydrobiol.* **52**, 1–184.

HUSMANN, S. (1958) Sand- und Schotterufer als Grenzbereiche limnologischer und bodenbiologischer Forschung, *Gewässer und Abwässer* **22**, 66–69.

HUSMANN, S. (1959) Neue Ergebnisse der Grundwasserbiologie und ihre Bedeutung für die Praxis der Trinkwasserversorgung, *Gewässer und Abwässer* **24**, 33–48.

JAKOBI, H. (1954) Biologie, Entwicklungsgeschichte und Systematik von *Bathynella natans* Vejd, *Zool. Jb. Syst.* **83**, 1–62.

JEANNEL, R. (1926) Faune cavernicole de la France, *Encycl. Entomol.* **VII**, Lechevalier, Paris, pp. 334.

KIEFER, F. (1957) Ruderfusskrebse (Crustacea, Copepoda) aus dem Grundwasser des südlichen Oberrheingebiets, *Mitt. Bad. Landesver. Naturkunde u. Naturschutz*, N. F. **7**, 53–68.

KIEFER, F. (1959) Unterirdisch lebende Ruderfusskrebse vom Hochrhein und Bodensee, *Beitr. z. naturkundl. Forsch. in SW-Deutschl.* **18**, 42–52.

LERUTH, R. (1939) La Biologie du domaine souterraine de la faune cavernicole de la Belgique, *Mém. Mus. Roy. H.N. Belg.* **87**, 1–506.

LÖFFLER, H. (1960) Die Entomostrakenfauna der Ziehbrunnen und einiger Quellen des nördlichen Burgenlandes, *Wiss. Arb. a. d.* Burgenland Eisenstadt **24**, 1–32.

LÖFFLER, H. (1960) 2. Beitrag zur Kenntnis der Entomostrakenfauna burgenländischer Brunnen und Quellen, ibid. **26**, 1–15.

MESTROV, M. (1960) Faunistisch-ökologische und biocönologische Untersuchungen unterirdischer Gewässer des Savetales, *Bioloski Glasnik* **13**, 73–109.

MOTAS, C. (1962) Procédés des sondages phréatiques—Division du domaine souterrain—Classification écologique des animaux souterrains—Le psammon, *Acta Mus. Mac. Sci. Nat. Skopje* **8**, 135–73.

MOTAS, C. (1962) Sur les Acariens phréatiques, leur distribution géographique et leur origine, *Zool. Anz.* **168**, 325–50.

MOTAS, C. (1963) On a recent report concerning the so-called hyporheic fauna, *Rev. Biol. (Bukarest)* **8**, 367–70.

MOTAS, C., TANASACHI, J., and ORGHIDAN, T. R. (1957) Über einige neue phreatische Hydrachnellae aus Rumänien und über Phreatobiologie, ein neues Kapitel der Limnologie, *Abh. naturw. Ver. Bremen* **35**, 101–22.

NOLL, W. (1939) Die Grundwasserfauna des Maingebietes, *Mitt. Nat. Mus. Aschaffenburg* **3**, 26, Jg. 1939.

NOLL, W. and STAMMER, H. J. (1953) Die Grundwasserfauna des Untermaingebietes von Hanau bis Würzburg mit Einschluss des Spessart, *Mitt. Naturwiss. Mus. Aschaffenburg*, N. F. **6**, 1–77.

ORGHIDAN, T. (1959) Ein neuer Lebensraum des unterirdischen Wassers: der hyporheische Biotop, *Arch. Hydrobiol.* **55**, 392–414.

PICARD, J. Y. (1962) Contribution à la connaissance de la faune psammique de Lorraine, *Vie et Milieu* **13**, 471–505.

REMANE, A. and SCHULZ, E, (1934) Das Küstengrundwasser als Lebensraum, *Schrift nat. Ver. Schlesw.-Holstein* **20**, 399–408.

RUFFO, S. (1961) Problemi relativi allo studio della fauna interstiziale iporreica, *Boll. di Zool.* **28**, 273–319.

RUTTNER-KOLISKO, A. (1956) Der Lebensraum des Limnopsammals, *Zool. Anz.*, Suppl., **14,** 421–7; weitere Arbeiten vgl. Litoral und Profundal sowie Fließwasserforschung.

SASSUCHIN, D. N. (1931) Die Lebensbedingungen der Mikrofauna in Sandanschwemmungen, *Arch. Hydrobiol.* **22,** 369–88.

SCHWOERBEL, J. (1961) Das unterirdische Wasser als Lebensraum, *Die Natur* **69,** 53–60; weitere Arbeiten siehe Fließwasserforschung.

SERBAN, M. (1963) La Récolte du matériel biologique des nappes phréatiques, *Acta Mus. Mac. Sci. Nat.* **9,** 1–13.

SPANDL, H. (1926) Die Tierwelt der unterirdischen Gewässer, *Speläologische Monographien* **11,** Vienna, pp. 235.

STAMMER, H. J. (1932) Die Fauna des Timavo. Ein Beitrag zur Kenntnis der Höhlengewässer des Süß- und Brackwassers im Karst, *Zool. Jb. Syst.* **63,** 521–656.

VANDEL, A. (1964) *Biopéologie. La biologie des animaux cavernicoles*, Gauthier-Villars, Paris, pp. 619.

VEJDOVSKI, T. (1882) *Tierische Organismen der Brunnengewässer von Prag*, Prague.

WISZNIEWSKI, J. (1934) Recherches écologiques sur le psammon, *Arch. Hydrobiol. Rybact.* **8,** 149–272.

WISZNIEWSKI, J. (1934) Remarques sur les conditions de la vie du psammon lacustre, *Verh. int. Ver. Limnol.* **6,** 263–74.

IX. RESEARCH INTO PRODUCTION

ARUGA, Y. (1966) Ecological studies of photosynthesis and matter production of Phytoplankton, III, Relationship between chlorophyll amount in water and primary productivity, *Bot. Mag. Tokyo* **79,** 20–27.

BIRGE, E. A. and JUDAY, C. (1922) The Inland Lakes of Wisconsin. The plankton. I. Its quantity and chemica composition, *Wisc. Geol. Nat. Hist. Surv.*, Bull. **64,** 1–222.

DAVIS, C. C. (1963) On questions of production and productivities in ecology, *Arch. Hydrobiol.* **53,** 145–61.

DOTY, M. S. and OGURI, M. (1959). The carbon fourteen technique for determining primary plankton productivity, *Publ. Staz. Zool. Napoli*, **31,** 70–94.

ECKSTEIN, H. (1964) Untersuchungen über den Einfluß des Rheinwassers auf die Limnologie des Schluchsees, *Arch. Hydrobiol.*, Suppl., **28,** 47–182.

EDMONDSON, W. T. (1960) Reproductive rates of rotifers in natural populations, *Mem. Ist. Ital. Idrobiol.* **12,** 21–77.

EDMONDSON, W. T. (1961) Secondary production and decomposition, *Verh. int. Ver. Limnol.* **14,** 316–39.

EDMONDSON, W. T. (1962) Food supply and reproduction of zooplankton in relation to phytoplankton population, *Rapp. Proc.-verb. Cons. perm. int. Explor. Mer* **153,** 137–141.

EDMONDSON, W. T. (1965) Reproductive rate of planctonic rotifers as related to food and temperature in nature, *Ecol. Monogr.* **35,** 61–111.

EICHHORN, R. (1957) Die Populationsdynamik der calanoiden Copepoden in Titisee und Feldsee, *Arch. Hydrobiol.*, Suppl., **24,** 186–246.

ELSTER, H.-J. (1954) Einige Gedanken zur Systematik, Zielsetzung und Terminologie der dynamischen Limnologie, *Arch. Hydrobiol.*, Suppl., **20,** 487–523.

ELSTER, H.-J. (1954) Über die Populationsdynamik von *Eudiaptomus gracilis* SARS und *Heterocope borealis* FISCHER im Bodensee-Obersee, *Arch. Hydrobiol.*, Suppl., **20,** 546–614.

ELSTER, H.-J. (1963) Die Stoffwechseldynamik der Binnengewässer, *Zool. Anz.*, Suppl., **27,** 335–87.

ELSTER, H.-J. and MOTSCH, B. (1966) Untersuchungen über das Phytoplankton und die organische Urproduktion in einigen Seen des Hochschwarzwaldes, im Schleinsee und Bodensee, *Arch. Hydrobiol.*, Suppl., **28,** 291–376.

ELSTER, H.-J. and VOLLENWEIDER, R. (1961) Beiträge zur Limnologie Ägyptens, *Arch. Hydrobiol.* **57,** 241–343.

FROLANDER, H. F. (1956) A plankton volume indicator, *J. Cons. int. Explor. Mer* **22,** 278–83.

GOLDMANN, C. R. (Ed.) (1966) *Primary Productivity in Aquatic Environments*, Univ. of California Press, Berkeley, pp. 464.

JØRGENSEN, C. B. (1966) *Biology of Suspension Feeding*, Pergamon Press, Oxford.

KALLE, K. (1945) *Der Stoffhaushalt des Meeres*. Akad. Verlagsges. Leipzig, 2nd ed., pp. 263.

KALLE, K. (1951). Die Mikrobestimmungen des Chlorophylls und der Eigenfluoreszenz des Meerwassers, *Deutsch. Hydrogr. Z.* **4,** 92–96.

KRAUSE, H. R. (1964) Zur Chemie und Biologie der Zersetzung von Süßwasserorganismen, unter besonderer Berücksichtigung des Abbaues der organischen Phosphorkomponenten, *Verh. int. Ver. Limnol.* **15,** 549–62.

LILLELUND, K. and KINZER, J. (1966) Absetz- und Verdrängungsvolumen von Planktonproben. Untersuchungen zur Methodik, *Int. Rev. Hydrobiol.* **51,** 757–74.

MARSHALL, S. M. and ORR, A. P. (1955) Experimental feeding of the copepod *Calanus finmarchicus* Gunner) on phytoplankton cultures labelled with radioactive carbon (C^{14}), *Deep Sea Res.* **3,** 110–14.

NAUWERCK, A. (1959) Zur Bestimmung der Filtrierrate limnischer Planktontiere, *Arch. Hydrobiol.*, Suppl., **25,** 83–101.

NAUWERCK, A. (1963) Die Beziehungen zwischen Zooplankton und Phytoplankton im See Erken, *Symb. Bot. Upsal.* **XVII, 5,** 1–163.

OHLE, W. (1956) Bioactivity, production and energy utilization of lakes, *Limnol. Oceanogr.* **1,** 139–49.

OSMERA, ST. (1966) Zur Methode der Festsetzung einiger Planktonkrebstiere, *Zool. Listy* **15,** 79–83.

PARSONS, T. R. and STRICKLAND, J. D. H. (1963) Discussion of spectrophotometric determination of marine plant pigments, with revised equations for ascertaining chlorophylls and carotenoids, *J. Mar. Res.* **21,** 155–63.

PRATT, D. M. and BERKSON, H. (1959) Two sources of error in the oxygen light and dark bottle method, *Limnol. Oceanogr.* **4,** 328–34.

RICHARDS, F. A. and THOMPSON, T. G. (1952) The estimation and characterisation of plankton populations by pigment analysis, II, A spectrophotometric method for the estimation of plankton pigments, *J. Mar. Biol.* **11,** 156–62.

RICHMANN, S. (1958) The transformation of energy by *Daphnia pulex*, *Ecol. Monogr.* **28,** 273–91.

RIGLER, F. H. (1961) The uptake and release of inorganic phosphorus by *Daphnia magna.*

RODHE, W. (1958) Primärproduktion und Seentypen, *Verh. int. Ver. Limnol.* **13,** 121–41.

RODHE, W. (1958) The primary production in lakes: some results and restrictions of the C^{14}-method, *Rapp. Proc.-verb. Cons. perm. int. Explor. Mer* **144,** 122–8.

RODHE, W. (1961) Die Dynamik des limnischen Stoff- und Energiehaushaltes, *Verh. int. Ver. Limnol.* **14,** 300–15.

RODINA, A. G. (1940) Bacteria and yeasts as food for cladocera *(Daphnia magna)*, *Compt. rend. Acad. Sci. USSR* **29,** 248–52.

RODINA, A. G. (1958) Mikroorganismen und die Steigerung der Fischproduktion der Gewässer, *Verlag der Akad. d. Wiss. USSR* (in Russian).

ŠESTÁK, Z. (1958) Quantitative determination of chlorophyll in algae, *Preslia* **30,** 138–45 (in Czech).

ŠESTÁK, Z. (1958) Methods of determining organic carbon in water and their use in the hydrobiology, *Acta Univ. Carol. Biologica*, Jg. **1958,** 269–81.

ŠESTÁK, Z. (1958) Paper chromatography of chloroplast pigments, *J. Chromatography* **1,** 293–308.

SOROKIN, J. I. (1966) Carbon-14-method in the study of the nutrition of aquatic animals, *Int. Rev. Hydrobiol.* **51,** 209–24.

SOROKIN, J. I. and PANOW, D. A. (1966) The use of C^{14} for the quantitative study of the nutrition of fish larvae, *Int. Rev. Hydrobiol.* **51,** 743–56.

SOUNDERS, G. W., TRAMA, F. B., and BACHMANN, R. W. (1962) Evaluation of a modified C^{14} technique for shipboard estimation of photosynthesis in large lakes, *Great Lakes Res. Div. Univ. Michigan* **8,** 1–61.

STEEMANN NIELSEN, E. (1952) The use of radio-active carbon (C^{14}) for measuring organic production in the sea, *J. Cons. int. Explor. Mer* **18,** 117–40.

STEEMANN NIELSEN, E. (1955) The production of organic matter by the phytoplankton in a Danish lake receiving extraordinarily great amounts of nutrient salts, *Hydrobiologia* **7,** 68–74.

STEEMANN NIELSEN, E. (1959) Untersuchungen über die primärproduktion des planktons in einigen Alpenseen Österreichs, *Oikos* **10,** 24–37.

STEEMANN NIELSEN, E. and AL KHOLY, A. A. (1956) Use of the C^{14} technique in measuring photosynthesis of phosphorus or nitrogen deficient algae, *Physiol. plantarum* **9,** 144.

STEEMANN NIELSEN, E. and HANSEN, V. K. (1959) Measurements with the carbon-14 technique of rates of respiration in natural populations of phytoplankton, *Deep Sea Res.* **5,** 222–33.

STRAUS, *Limnol. Oceanogr.* **6,** 165–74.

STRICKLAND, J. D. H. (1960) Measuring the production of marine phytoplankton, *Fish. Res. Board Canada* **122,** 1–172.

SUTCLIFFE, W. H. (1957) An improved method for the determination of preserved plankton volumes, *Limnol. Oceanogr.* **2,** 295–6.

TALLING, J. F. (1957) Diurnal changes of stratification and photosynthesis in some tropical African waters, *Proc. Roy. Soc. London*, B, **147,** 57–83.

THIENEMANN, A. (1931) Der Produktionbegriff in der Biologie, *Arch. Hydrobiol.* **22,** 616–22.

TRANTER, D. J. (1965) A method for determining zooplankton volumes, *J. Cons. int. Explor. Mer* **25,** 272–8.

VOLLENWEIDER, R. A. (1965) Calculation models of photosynthesis-depth curves and some implications regarding day rate estimates in primary production measurements, *Mem. Ist. Ital. Idrobiol.*, Suppl., **18.**

WATT, W. D. (1966) Release of dissolved organic material from the cells of phytoplankton populations, *Proc. Roy. Soc.*, B, **164,** 521–51.

WETZEL, R. G. (1965) Techniques and problems of primary productivity measurements in higher plants and periphyton, *Mem. Ist. Ital. Idrobiol.*, Suppl., **18.**

YENTSCH, C. J. R. and HEBARD, F. J. (1956) A gauge for determining plankton volume by the mercury immersion method, *J. Cons. int. Explor. Mer* **22,** 184–90.

X. CULTURE METHODS

BLUNCK, H. (1924) Die Zucht der Wasserkäfer, ABDERHALDEN **IX, 2,** 293–310.

CHU, S. P. (1942) The influence of the mineral composition of the medium on the growth of planktonic algae. I, Methods and culture media, *J. Ecol.* **30,** 284–325.

DROOP, M. (1962) Organic micronutrients, LEWIN, R. A. (ed.) in *Physiology and Biochemistry of Algae*, Academic Press.

FLÜCHINGER, E. and FLUCH, H. (1949) Ein künstliches Milieu für die Zuchten von *Daphnien* im Laboratorium, *Experientia* **5,** 486.

FOTT, B. and TRUNCOVA, E. (1964) List of species in the culture collection of algae at the Department of Botany of Charles University, *Acta Univ. Carol. Biologica*, **2,** 97–110.

GALTSOFF, P. S. (1937) *Culture Methods for Invertebrate Animals*, Ithaca, New York, pp. 590.

JAMES, T. W. (1960) Uptake and localization of thioglycolate in the synchronized flagellate *Astasia longa*, *10 Congr Int. Biol. Cellul.* **9,** 510–14.

KOCH, W. (1964) Artenliste der Algensammlungen am Pflanzenphysiologischen Institut der Universität Göttingen, *Arch. Mikrobiol.* **48.**

LÉPINEY, L. DE (1962) Sur l'élevage de Copépodes au laboratoire, *Hydrobiologia* **20,** 217–22.

LORENZEN, H. (1959) Die photosynthetische Sauerstoffproduktion wachsender *Chlorella* bei langfristig intermittierender Belichtung, *Flora* **147,** 382–404.

LORENZEN, H. (1964) Synchronization of *Chlorella* with light-dark changes and periodical dilution to a standard cell number, in ZEUTHEN, E. (ed.) *Synchrony in Cell Division and Growth*, Interscience, New York.

MYERS, J. (1962) Laboratory cultures, in LEWIN, R. A. (ed.) *Physiology and Biochemistry of Algae*, Academic Press, New York, 603–16.

NAUMANN, E. (1929) Die Zucht des Phytoplanktons, ABDERHALDEN **IX, 2,** 1424–34.

NAUMANN, E. (1929) Die Zucht der Cladoceren des Seeplanktons, ABDERHALDEN **IX, 2,** 1435–42.

NAUMANN, E. (1929) Die Zucht einiger Cladoceren des Litorals, ABDERHALDEN **IX, 2,** 1443–50.

NAUMANN, E. (1929) Die Zucht einiger Cladoceren des Teichplanktons, ABDERHALDEN **IX, 2,** 1451–62.

NAUMANN, E. (1929) Über die Zucht eines Ostracoden, *Cyprinotus incongruens* Rauch, ABDERHALDEN **IX, 2,** 1463–65.

NAUMANN, E. (1929) Die Zucht der Copepoden des Seeplanktons, ABDERHALDEN **IX, 2,** 1466–9.

NAUMANN, E. (1929) Die Zucht planktischer Rotatorien, ABDERHALDEN **IX, 2,** 1471–4.

NAUMANN, E. (1929) Die Zucht Aufwuches unter Anwendung der Methode des Aufwuchsträgers, ABDERHALDEN **IX, 2,** 1475–80.

PIRSON, A. (1962) Synchronisation von *Chlorella* durch Licht-Dunkel-Wechsel, *Vortr. Gesamtgeb. Bot.*, N. F. **1,** 178–86.

PRINGSHEIM, E. G. (1946) *Pure Cultures of Algae*, Cambridge Univ. Press.

PROVASOLI, L. (1960) Artificial media for freshwater algae: problems and suggestions, in *The Ecology of Algae*, **2,** 84–96, Pittsburgh Univ. Press.

RODHE, W. (1948) Environmental requirements of fresh-water plankton algae, *Symb. Bot. Upsal.* **10,** 1, 1–149.

RYLOV, W. M. (1929) Haltung und Aufzucht von Süßwasserbryozoen, ABDERHALDEN **IX, 2,** 1419–23.

RYTHER, J. H. (1954) Inhibitory effects of phytoplankton upon the feeding of *Daphnia magna* with reference to growth, reproduction and survival, *Ecology* **35,** 522–33.

SENGER, H. (1964) Eine automatische Verdünnungsanlage und ihre Anwendung zur Erzielung homokontinuierlicher *Chlorella*-Kulturen, *Arch. Mikrobiol.* **48,** 91–94.

SOEDER, C. J. and RIED, A. (1963) Über die Atmung synchron kultivierter *Chlorella*. I, Veränderungen des respiratorischen Gaswechsels im Laufe des Entwicklungscyclus, *Arch. Mikrobiol.* **45,** 343–58.

SPOEHR, H. A. and MILNER, H. W. (1949) The chemical composition of *Chlorella*: effect of environmental conditions, *Plant Physiol.* **24,** 124–49.

STARR, R. C. (1960) The culture collection of algae at Indiana University, *Amer. J. Bot.* **47**, 67–86.
STEINER, G. (1963) *Das Zoologische Laboratorium*, Schweizerbart, Stuttgart.
THIENEMANN, A. (1924) Die Zucht der Dipteren und Wasserhymenopteren, ABDERHALDEN **IX, 2,** 311–18.
TÜMPEL, R. (1924) Die Zucht der Odonaten, ABDERHALDEN **IX, 2,** 285–6.
ULMER, G. (1924) Zucht der Trichoptera (Köcherfliegen), Lepidoptera (Schmetterlinge), Ephemeroptera (Eintagsfliegen), Plecoptera (Steinfliegen), ABDERHALDEN **IX, 2,** 287–91.
WACHS, B. (1965) Vorkommen und Verbreitung der Oligochaeten in der Edertalsperre, *Arch. Hydrobiol.* **61,** 190–204.
WAGLER, E. (1924) Zucht von Krebsen und Würmern, ABDERHALDEN **IX, 2,** 319–64.
WELLS, M. M. (1932) *The Collection and Preservation of Animal Forms*, Chicago, Ill., pp. 72.
WITSCH, H. v. (1950) Physiologischer Zustand und Wachstumsintensität bei *Chlorella, Arch. Mikrobiol.* **14,** 128–41.

XI. WORKS FOR IDENTIFICATION

1. PLANTS

FOTT, B. (1959) *Algenkunde*, G. Fischer, Jena, pp. 482.
HUBER-PESTALOZZI, G. (1938–61) Das phytoplankton des Süsswassers, Parts 1–5. *Die Binnengewässer* **16.**
HUSTEDT, F. (1924) Vom Sammeln und Präparieren der Kieselalgen sowie Anagaben über Untersuchungs- und Kulturmethoden, ABDERHALDEN **XI, 4.**
HUSTEDT, F. (1961) *Kieselalgen (Diatomeen)*, Franckh, Stuttgart, 2nd edn. pp. 70.
ISLAM, N. (1963) A revision of the genus *Stigeoclonium*, Cramer, Weinheim, pp. 164.
KLOTTER, H. E. (1959) *Grünalgen (Chlorophyceen)*, Franckh, Stuttgart, pp. 76.
KRIEGER, W. and GERLOFF, J. (1962) *Die Gattung* Cosmarium, No. 1, Cramer, Weinheim, pp. 112.
LINDAU, G. (1914–16) *Kryptogamenflora für Anfänger*, Vol. IV, sects. 1–3, Springer, Berlin, pp. 219, 200, and 125; sect. 1 revised by Melchior, H. (1926) pp. 314.
MAUCH, E. (1966) *Bestimmungsliteratur für Wasserorganismen im mitteleuropäischen Gebiet*, G. Fischer, Stuttgart, pp. 22.
PASCHER, A. (1913–1932) *Die Süsswasserflora Deutschlands, Österreichs und der Schweiz*, Nos. I–XII, G. Fischer, Jena.
PRINTZ, H. (1964) Die Chaetophoralen der Binnengewässer, eine systematische Übersicht, *Hydrobiologia* **24,** 1–376.
RABENHORST, L. (1930) *Kryptogamenflora von Deutschland, Österreich und der Schweiz*, vol. X–XIV, 2nd edn., Kolkwitz, R. (ed.), Akad. Verlagsges., Leipzig.
RIETH, A. (1961) *Jochalgen (Konjugaten)*, Franckh, Stuttgart, pp. 86.
SCHÖMMER, F. (1949) *Kryptogamen-Praktikum*, Franckh, Stuttgart, pp. 492.
STEINECKE, F. (1958) Das plankton des Süsswassers, *Biolog. Arbeitsbücher* **1,** Quelle & Meyer, Heidelburg, pp. 71.
SKUJA, H. (1948) Taxonomie des Phytoplanktons einiger Seen in Uppland, Schweden, *Symb. Bot. Upsal.* **IX,** 3, 1–399.
SKUJA, H. (1956) Taxonomische und biologische Studien über das Phytoplankton schwedischer Binnengewässer, *Nova Acta Reg. Soc. Sci. Upsal.*, Ser. IV, **16,** 1–404.

2. ANIMALS

ARNDT, W. (1928) Porfera, Schwämme, Spongien, DAHL **4,** 1–94.
BRAUER, A. (1909 ff.) *Die Süsswasserfauna Deutschlands*, G. Fischer, Jena, reprinted 1961 onward.
BROCH, HJ. (1928) Hydrozoen, DAHL, **4,** 95–160.
BROHMER, A. (1964) *Fauna von Deutschland*, 9th edn., Quelle & Meyer, Heidelberg.
BROHMER, EHRMANN, ULMER (1927 ff.) *Die Tierwelt Mitteleuropas*, Quelle & Meyer, (hereafter referred to as BROHMER).
DAHL, FR. (1925 ff.) *Die Tierwelt Deutschlands und der angrenzenden Meeresteile*, G. Fischer, Jena, (hereafter referred to as DAHL).
DONNER, J. (1962) *Rädertiere (Rotatoria)*, Franckh, Stuttgart, pp. 50.
EHRMANN, P. (1956) *Weichtiere, Molluska*, BROHMER, vol. **II,** sect. **I,** 1–264, and the supplement by Zilch, A. and Jaeckel, S. G. A. (1962).
ENGELHARDT, W. (1959) *Was lebt im Tümpel, Bach und Weiher?*, Franckh, Stuttgart, pp. 258.

EYFERTH, B. and SCHOENICHEN, W. (1925) *Einfachste Lebensformen des Tier- und Pflanzenreiches*, 5th edn., Berlin, (op).
GROSPIETSCH, TH. (1958) *Wechseltierchen (Rhizopoda)*, Franckh, Stuttgart, pp. 82.
HERBST, H. V. (1962) *Blattfusskrebse (Phyllopoda)*, Franckh, Stuttgart, pp. 129.
ILLIES, J. (1955) Steinfliegen oder Plecoptera, DAHL **43**, 1–150.
JOHANNSON, L. (1929) Hirudinea (Egel), DAHL **15**, 133–55.
KAHL, A. (1930–1935) Urtiere oder Protozoa, I, Wimpertiere oder Ciliata (infusoria), DAHL **18, 21, 25, 30,** 1–886.
KIEFER, F. (1960) *Ruderfusskrebse (Copepoda)*, Franckh, Stuttgart, pp. 97.
KLIE, W. (1938) Muschelkrebse (Ostracoda), DAHL **34**, 1–230.
LANG, A. (1948) *Monographie der Harpacticida*, Lund, Sweden.
LUTHER, A. (1955) Die Dalyelliden (Turbellaria, Neorhabdocoela), *Acta Zool. Fenn.* **87**, pp. 337.
MACAN, T. T. (1955) A key to the nymphs of British Ephemeroptera, *The Salmon Trout Mag.*, Jan. 1955, 79–90.
MACAN, T. T. (1956) A revised key to the British water bugs *(Hemiptera-Heteroptera)*, *Freshw. Biol. Ass. Sci. Publ.* **16**, 1–73.
MANN, K. H. (1954) A key to the British freshwater leeches with notes on their ecology, *Freshw. Biol. Ass. Sci. Publ.* **14**.
MARCUS, E. (1925) *Bryozoa. Biol. Tiere Deutschl.* **47**, 1–46.
MARCUS, E. (1928) Bärtierchen (Tardigrada), DAHL **2**, 1–230.
MAY, E. 1933. Odonata, DAHL **27**, 1–124.
MEYL, A. (1961) *Fadenwürmer (Nematoda)*, Franckh, Stuttgart, pp. 74.
RAMAZZOTTI, G. (1963) Tardigrada, *Mem. Ist. Ital. Idrobiol.*, Suppl.
REMANE, A. (1935–6) Gastrotricha, *Kl. u. Ordn. Tierr.* **4**, sect. 2, book 1, part 2, 1–242.
RYLOV, W. M. (1935) Das Zooplankton des Süsswassers, *Die Binnengewässer* **15**, Schweizerbart, pp. 272.
SCHELLENBERG, A. (1928) Zehnfüßer (Decapoda), DAHL **10**, 1–146.
SCHELLENBERG, A. (1941) Flohkrebse (Amphipoda), DAHL **40**, 1–252.
SCHNEIDER, W. (1939) Fadenwürmer (Nematodes), DAHL **36**, 1–260.
SCHOENEMUND, E. (1930) Eintagsfliegen oder Ephemeroptera, DAHL **19**, 1–106.
SCOURFIELD, D. J. and HARDING, J. P. (1958) A key to the British species of freshwater Cladocera, *Freshw. Biol. Ass. Sci. Publ.* **5**, pp. 55.
SIEWING, R. (1959) Syncarida, *Kl. u. Ordn. Tierr.* **5**, sect. 1, book 4, part 2, 1–121.
SPERBER, C. (1948) The taxonomical study of the Naidide, *Zool. Bidr. Uppsala* **28**, 1–296.
STRESEMANN, E. *Exkursionfauna von Deutschland*, Volk and Wissen, Berlin; I, *Wirbellose (ausser insekten)* (1957), pp. 488; II, *Wirbellose Insekten 1* (1964) pp. 518, *Insekten 2* (to be published); III, *Wirbeltiere*, 2nd edn., pp. 352.
UDE, H. (1929) Oligochaeta, DAHL **15**, 1–132.
ULMER, G. (1925) Trichoptera, *Biol. Tiere Deutschl.* **13**, 1–113.
VIETS, K. (1936) Wassermilben (Hydracarinen), DAHL **32**, 1–574.
VIETS, K. and VIETS, K. O. (1960) supplement to Viets, Wassermilben (Hydracarina) in BROHMER **I**, 1–44.
VOIGT, M. (1957) *Rotatoria. Die Rädertiere Mitteleuropas*, Bornträger, Berlin, vol. 2.
ZILCH, A. and JAECKEL, S. G. A. (1962) supplement to Ehrmann (1956), in BROHMER.

XII. BIOLOGICAL ESTIMATION OF WATER QUALITY

BEER, W.-D. (1958) Zur Problematik des biologischen Gütelängsschnittes von Fliessgewässern, dargestellt am Beispiel der Weissen Elster, *Wasserwirtschaft-Wassertechnik* **8**, 195–9.
BEER, W. D. (1961) Methodologische Untersuchungen zur biologischen Fliessgewässeranalyse, *Int. Rev. Hydrobiol.* **46**, 5–17.
BEER, W.-D. (1962) Zur Methodik und Problematik des biologischen Flussgütelängsschnittes, *Wiss. Z. Univ. Leipzig, math.-nat.* **11**, 143–8.
BEER. W.-D. (1964) Organismen als Anzeiger der Wassergüte und Saprobiensystem, *Wiss. Z. Univ. Leipzig, math.-nath.* **13**, 25–30.
BERGER, H. (1966) *Leitfaden der Trink- und Brauchwasserbiologie*, 2nd edn., Gerloff, J. and Lüdemann (eds.), G. Fischer, Stuttgart, pp. 360.
BICK, H. (1963) A review of central European methods for the biological estimation of water pollution levels, *Bull. Wld. Hlth. Org.* **29**, 401–13.

REFERENCES

BREITIG, G. (1961) Vorschlag einer Einheitsmethodik zur biologischen Untersuchung von Fliessgewässern, *Mitt. Inst. Wasserwirtsch. Berlin* **21,** 99–118.

BRINGMANN, G. and KÜHN, R. (1958) Veränderung der Eutrophierung und Bioproduktion gemessen am Biomassentiter von Testalgen, *Ges.-Ing.* **79,** 50–54.

BRINGMANN, G. and KÜHN, R. (1958) Der Biomassentiter von Spaltalgen als Massstab der Belastung des Wassers durch nichtmineralisierten Stickstoff, *Ges.-Ing.* **79,** 329–33.

BRINGMANN, G. and KÜHN, R. (1959) Vergleichende wassertoxikologische Untersuchungen an Bakterien, Algen und Kleinkrebsen, *Ges.-Ing.* **80,** 115–20.

BRINGMANN, G. and KÜHN, R. (1962) Biomassentiter und Saprobien—eine hydrobiologische Vergleichsanalyse an Niederrhein, Fulda und Havel, *Int. Rev. Hydrobiol.* **47,** 123–45.

BRINGMANN, G., KÜHN, R., and LÜDEMANN, D. (1962) Bedeutung und Zielsetzung biologischer Wasseruntersuchungen, *Gwf* **103,** 1127–32 and 1232–6.

BRUCKSTEEG, W. and THIELE, H. (1959) Die Beurteilung von Abwasser und Schlamm mittels TTC (2,3,5-Triphenyltetrazoliumchlorid), *Gwf* **100,** 916–20.

CASPERS, H. (1962) Die Bestimmung der Sedimentaktivität, *Int. Rev. Hydrobiol.* **47,** 581–6.

CASPERS, H. and KARBE, L. (1965) Trophie und Saprobie als stoffwechseldynamischer Komplex, *Arch. Hydrobiol.* **61,** 453–70.

CASPERS, H. and SCHULZ, H. (1960) Studien zur Wertung der Saprobiensysteme. Erfahrungen an einem Stadtkanal Hamburgs, *Int. Rev. Hydrobiol.* **45,** 535–65.

CASPERS, H. and SCHULZ, H. (1962) Weitere Unterlagen zur Prüfung der Saprobiensysteme, *Int. Rev. Hydrobiol.* **47,** 100–17.

COHN, F. (1853) Über lebende Organismen im Trinkwasser, *Z. klin. Med.* **4,** 229–37.

ELSTER, H.-J. (1962) Seetypen, Fließgewässertypen und Saprobiensysteme, *Int. Rev. Hydrobiol.* **47,** 211–18.

ELSTER, H.-J. (1966) Über die limnologischen Grundlagen der biologischen Gewässer-Beurteilung in Mitteleuropa, *Verh. int. Ver. Limnol.* **16,** 759–85.

FJERDINGSTAD, E. (1964) Pollution of streams estimated by benthonic phytomicro-organisms: I, A saprobic system based on communities of organisms and ecological factors, *Int. Rev. Hydrobiol.* **49,** 63–131.

GABRIEL, J. (1946) Principy biologichého hodnoceni vody. *Čas Lek. ces.* **85,** 1425–31.

HORASAWA, I. (1956) A preliminary report on the biological index of water pollution, *Zool. Mag. (Tokyo)* **54,** 1.

KNÖPP, H. (1955) Grundsätzliches zur Frage biologischer Vorfluteruntersuchungen, erläutert an einem Gütelängsschnitt des Mains, *Arch. Hydrobiol.*, Suppl., **22,** 363–8.

KNÖPP, H. (1961) Der AZ-Test, ein neues Verfahren zur toxikologischen Prüfung von Abwässern (Begründung und Beschreibung der Methode), *Dtsch. Gewässerkdl. Mitt.* **5,** 66–73.

KNÖPP, H. (1964) Die "Zusätzliche Zehrung"—eine neue biochemische Kennzahl zur Bewertung von Selbstreinigungskraft und Verschmutzung, *Gwf* **105,** Sep. pp. 6.

KOLKWITZ, R. (1950) Ökologie der Saprobien, *Schriftenreihe Ver. Wasser-, Boden-, Lufthygiene* **4,** Piscator, Stuttgart.

KOLKWITZ, R. and MARSSON, M. (1902) Grundsätzliches für die biologische Beurteilung des Wassers nach seiner Flora und Fauna, *Mitt. K. Prüfanst. Wasserversorg. Abwasserbes. Berlin-Dahlem* **1,** 33.

KOLKWITZ, R. and MARSSON, M. (1908) Ökologie der pflanzlichen Saprobien. *Ber. dt. bot. Ges.* **26a,** 505–19.

KOLKWITZ, R. and MARSSON, M. (1909) Ökologie der tierischen Saprobien, *Int. Rev. Hydrobiol.* **2,** 126–52.

KOTHÉ, P. (1962) Der "Artenfehlbetrag", ein einfaches Gütekriterium und seine Anwendung bei biologischen Vorflutuntersuchungen, *Dt. Gewässerkdl. Mitt.* **6,** 60–65.

LAUTERBORN, R. (1901) Die sapropelische Lebewelt, *Zool. Anz.* **24,** 50–55.

LAUTERBORN, R. (1903) *Die Verunreinigung der Gewässer und die biologische Methode ihrer Untersuchung,* Ludwigshafen, pp. 33.

LIEBMANN, H. (1947) Die Notwendigkeit einer Revision des Saprobiensystems und deren Bedeutung für die Wasserbeurteilung, *Ges.-Ing.* **68,** 33–37.

LIEBMANN, H. (1955) Die Kartierung der Wassergüte, beschrieben an Flußstauen und Seen Süddeutschlands, *Ber. Abwassertech. Vereinig.* **II,** 62–69.

LIEBMANN, H. (1959) Methodik und Auswertung der biologischen Wassergütekartierung. *Münch. Beitr. Abwasser-, Fischerei- und Flußbiologie* **6,** 143–56.

LIEBMANN, H. (1959) Folgerungen für die Praxis aus den Ausführungen über die Bewertung der Wasserqualität, *Münch. Beitr.* **6,** 181–5.

LIEBMANN, H. (1962) *Handbuch der Frischwasser- und Abwasserbiologie,* Vol. II, 2nd. edn., Oldenbourg, Munich, pp, 588.

LUND, J. W. G. (1960) The microscopical examination of freshwater, *Proc. Soc. Water treatment Examination* **9,** 109–44.

MADLER, K. (1964) Die natürliche Selbstreinigung in fließenden Gewässern. *Wiss. Z. Univ. Leipzig, math. nat.* **13,** 13–16.

MAUCH, E. (1963) Untersuchungen über das Benthos der deutschen Mosel unter besonderer Berücksichtigung der Wassergüte, *Mitt. Zool. Mus. Berlin* **39,** 1–172.

MEZ, C. (1898) *Mikroskopische Trinkwasseranalyse*, Springer, Berlin.

PANTLE, R. and BUCK, H. (1955) Die biologische Überwachung der Gewässer und die Darstellung der Ergebnisse, *Gwf* **96,** 604.

SCHRÄDER, T. (1959) Zur Limnologie und Abwasserbiologie von Talsperren Obere Saale (Thüringen), *Int. Rev. Hydrobiol.* **44,** 485–619.

SLÁDEČEK, V. (1961) Zur biologischen Gliederung der höheren Saprobitätsstufen, *Arch. Hydrobiol.* **58,** 103–21.

SLÁDEČEK, V. (1963) A guide to limnosaprobical organisms, *Sci. Pap. Inst. Techn. Prag.*, *Technol. Water* **7,** 543–612.

SLÁDEČKOVÁ, A. and SLÁDEČEK, V. (1963) Periphyton as indicator of the reservoir water quality. I, True periphyton, *Sci. Pap. Inst. Chem. Techn. Prag. Technol. Water* **7,** 507–61.

ŠRÁMEK-HUŠEK, R. (1956) Zur Charakteristik der Höheren Saprobitätsstufen, *Arch. Hydrobiol.* **51,** 376–90.

TÜMPLING, W. v. (1960) Probleme, Methoden und Ergebnisse biologischer Güteuntersuchungen an Vorflutern, dargestellt am Beispiel der Werra, *Int. Rev. Hydrobiol.* **45,** 513–34.

TÜMPLING, W. v. and ZIEMANN, H. (1962) Beitrag zur Kritik der biologischen Einheitsmethodik, *Mitt. Inst. Wasserwirtsch*, No. 12.

UHLMANN, D. (1964) Die biologische Selbstreinigung in Gewässern, *Wiss. Z. Univ. Leipzig, math. nat.* **13,** 17–24.

WETZEL, A. (1964) Der Anteil der Wasserorganismen and der Wassergüte, *Wiss. Z. Univ. Leipzig, math. nat.* **13,** 3–6.

ZAHNER, R. (1962) Über die Wirkung von Treibstoffen und Ölen auf Regenbogenforellen, *Vom Wasser* **29,** 152–77.

ZAHNER, R. (1964) Beziehungen zwischen dem Auftreten von Tubificiden und der Zufuhr organischer Stoffe im Bodensee, *Int. Rev. Hydrobiol.* **48,** 417–54.

ZELINKA, M. and MARVAN, P. (1961) Zur Präzisierung der biologischen Klassifikation der Reinheit fliessender Gewässer, *Arch. Hydrobiol.* **57,** 389–407.

INDEX

Page numbers in italics refer to illustration

Abiotic zone 150
Absorption maximum for chlorophyll 136
Abundance scale (for organisms) 151
Acanthodiaptomus denticornis 65
Aeolosoma 160
Agapetus *110*
Aggressive carbon dioxide 22
Air bubble 142
Alderflies (*Sialis*) 85
Algae 6, 70, 108, 141, 142
  crusts of, on stones 73, 106, 124
  cultures of 164, 167
  in flowing water 104, 106, 108, 124
  nutrition of zooplankton 167
  races of 165
Alpha-mesosaprobic zone 150
Alpha-polysaprobic zone 150
Amino acids 166
Ammonium compounds 28
Amphipoda 130
Ampoules for C^{14} standard solution 144
Ancylus fluviatilis *110*
Antisaprobic conditions 150
Aphanothece 80
Aquatic types 1
Arcella *67*
Artificial substrates for *Aufwuchs* 81, 124, 155
Ash weight 100, 111
Assimilated C^{14} 140, 141
Assimilation (photosynthetic) capacity 65, 137
Assimilation (photosynthesis) curve of phytoplankton *142*
Asterionella 52
Auerbach's box grab 88, 89
Aufwuchs 2, 76, 79, 81, 83, 142, 155
  see also Algae
Automatic method for sieving mud 95, 96

Bacterial decomposition 7, 134
  see also Decay; Putrefaction
Bacterial growth 138
Bait cans 130
Baits for cave animals 130
Barbel 105
Barbel region 105
Barrier-layer photo-cell 16, 17
Bdelloid rotifers 76
Benthal 69, 84, 102
Benthos 6, 104
Beta-mesosaprobic zone 149
Beta-polysaprobic zone 150
Biochemical oxygen demand (BOD) 158
Biocoenoses 2
Biogenic metabolic cycle 6, 134
Biological index of pollution 157
Biological longitudinal section of the quality of water 151, 152
Biological zonation in flowing waters 126
Biomass 83, 100, 106, 135
Biomass titre techniques 158
Biotin 166
Blue band filter 137
Blue–green algae 73, 83
Bolting cloths 37, 38, 168
Bosmina 66, 146
Botrydiopsis arhiza 67
Bottles for taking samples of water and plankton 11, 38, 45, 53, 57, 118
Bottles, shaking 49
Bottom deposits 80
Bottom fauna 104
  see also Benthos
Bottom grabs 74, 76, 85, 112, 117
Bottom organisms 4, 84
Bottom-skinner *93*
Bottom water bottle 92
Bottom zone of a lake 69, 84, 92
Bouin 160, 161
Bream region 105
Bronze gauze 117
Bryozoa 159, 162
Buffering of culture media 166

C^{14}, standard solution 143, 144
Caddis-fly larvae (preservation and fixation) 99, 111, 122, 161
Calcium hardness of water 30
Callitriche 106
Campanella umbellaria 83
Carbon assimilations of *Aufwuchs* algae 142
Carbon dioxide 6, 164, 167
  equilibrium 22
Carbon dioxides, firmly bound 30
Carbonate hardness 30
Carchesium polypinum 83

Catharobic zone 150
Cave-streams 1
Cave walls 130
  streaming with water 130
Cave waters 130
Cellafilter 136
Celluloid strips 81
Centrifugation 55
Ceratium 5, 52, 55, 56, 162
Ceratophyllum 75
Cerophyl 163
Chaoborus *37*, 85, 92, 95
Characeae 75, 80
Chelators 166
Chemoautotrophic plants 134, 141
Chemosynthesis 70, 134, 141, 145
Chironomid larvulae *127*
Chironomids 85, 102, 127
Chironomus (= *Tendipes*) 84, 95, 102
Chironomus lakes 85
Chlorella 162, 167
  nutrient solution for 165
Chlorophyll
  determinations of 65, 84, 136
  fluorescent light of 136
Chromogaster 162
Chromophyton rosanoffi 67
Chromosmium acetic acid 159
Chromulina 167
Chrysomonads 167
Circulation of water in lakes 3
Cladocera (culture) 163
Cladophora 80
Closing net 38, 42, 43, 52
Closterium *54*
Closure mechanism 43
Cocaine 160
Coccoid green algae 167
Coleochaete soluta 83
Coleoptera 99, 159
Collecting funnel 102, 103 (littoral), 122 (for rivers)
  Noll's 131, 132
Collection by hand 99
Collembola 159
Collodion 108
Colonization rate 125
Colour comparator 21
Coloured adhesive tape for sand tubes 120
Colourless algae 164
Compound chamber 59, 61
Condosiga (botrytis) *67*, 83
Consumers 2, 134, 145, 147, 157
Contact-zone between sediment and water 92
Conversion of temperature degrees 16
Copepodite 92
Copepods (culture) 52
Coprozoic zone 150
Cork stoppers 81
Counting
  of phytoplankton 61, 62
  of zooplankton 48, 49
Counting chamber 49
Counting plates 49
Crawling rate 1
Crenobia alpina 1
Crenothrix fusca 83
Cube root curves 65
Culture
  of algae 164
  of *Synura* 166
Culture glasses 163
Cups worn in rocks by dripping 130
Current, interrupted direct, for collection from bottom samples 99
Cyclops 65, 85, 163
Cyprideis litoralis 163

Dalyellidae 160
Danube 124
Daphnia, daphnias 4, 6, 52, 65, 72, 145, 158, 162
Daily periodic emergence of insects 103
Dark flasks 140
Daylight lamps 167
Day–night fluctuation of the drift 122
Day–night rhythm of cave animals 129
Dead water 104
Decapoda 129
Decay 134
Deep region 69
Deep sediment of lakes 88
Degeneration phase in algae 167
Depth distribution of phanerogams 70, 75
Dero 160
Detritus 115, 118, 124, 136
Developmental period of eggs 146
Developmental rhythm of cave animals 129
Dewar (thermos) bottles 73, 112
Diametric counting 62
Diaphanosoma *66*
Diaptomus 146
Diatoms 65, 76, 80, 83, 106, 124
Dinobryon divergens 166
Dividing grab of Lenz *87*, 88
Dredge
  closing, of Riedl 94
  of Usinger and Needham 115
Dredge-sledge of Elster *94*
Dredges 74, 75, 77–79, 93, 106, 112, 115, 118, 130
Drift (organismal) 121
Dripstones 130
Drop-bottle of Herbst *46*, 47
Dugesia gonocephala 1
Durognost T 31

Ecdyonorius 111
Echo-sounder 35, *36*

EDTA 166
Ekman–Birge grab 85, *86*, 87
Electrical centrifuge 64
Electrical thermometer resistance 15
Electrometric cell-counter 167
Electrometric determination of pH 21
Elodea 75, 80, 162
Entomostraca 76
Epeorus 111, 164
Ephemeroptera (may-flies) 99, 111, 122, 161
Epilimnion 3, 70
Epilithic algae 73, 106
Epineuston 67
Epiphanes senta 164
Erkenia 167
Erpobdellidae 162
Error, sources of
  in bottom grabs 85
  in flotation techniques 114
  in shovel-samplers 118
Escherichia 158
Ethylene diamine tetra-acetic acid 166
Eucricotopus-brevipalpis 80
Eudiaptomus gracilis 162
Eudiaptomus graciloides (filtration rate) 146
Euglena viridis 164
Eulittoral region 69, 70
Eupsammon 71
Eusaprobic conditions 156
Eutrophic lakes 84
Exposure
  of microscope slides 81, 82
  of sand tubes 120
Exposure period
  of artificial substances in flowing waters 124
  of C^{14} samples (no number given) 144, 145

Feeding rate of zooplankton 146
Filamentous algae 76, 80
Filter feeders 61, 145
Filtration rate
  dependence of, on the food supply 146
  of zooplankton 145
Filtration resistance 55
Filtration surfaces 37, 41, 85
Fishery biology ix
Fixation of zooplankton 48
Flight behaviour
  of plankton in front of a net 38, 48
  of species living in running water 73, 111
Flight period 122
Flight reaction of species living on stones 73
Floating leaves 2, 80
  plants with 70
Flotation techniques 113
Fluorescent tubes 167
Fontinalis 80, 106
Food chains 145
Foot candle 17 n.
Form•resistance 4
Formalin 83, 99, 101, 106, 107, 159
Fragillaria crotonensis *5*, 52
Fresh volume 83
Fresh weight (moist weight) 83, 100, 101, 106, 111, 126
Friedinger-bottle *46*, 47, 48, 92
Full circulation 84
Funnel trap *123*
Fusaria 106

Gamma-mesosaprobiotic zone 150
Gamma-polysaprobiotic zone 150
Gastrotricha 159
Geiger–Müller counter 64, 138, 140, 144
Gerridae 67
Gessner funnel for measuring the water current 18
Glass bottles (flasks) 10
Glass plates 82, 83, 155
Glass tube for periphyton 76
Glossiphonidae 162
Glyceria 106
Glycerophosphate 166
Glycylglycine 166
Gordiidae 159
Grabs 90, 91, 92
  Günther's 90, *91*
  Lang's 91
  Van Veen's *90*, 91
Gran method 137
Green algae 80, 83, 106
Ground shovel for streams 114, 115, 116, 117

Habrotrocha constricta 162
Haematoxylin (Delafield's) 107
Haemocytometer (counting chamber) 167
Halacaridae 130, 164
Hand net 75, 120, 130
Harpacticid as a cleaner 163
Harpacticids 85, 130, 163
Hehner cylinder 28
Helmet of Daphnias 6
Hemiptera 160
Heptagenia 111
Higher aquatic plants 106, 108, 149
Hildenbrandia 106, 149
Hirudinea 159, 162
Histidine 166
Historical deposits in lakes 93
Homocontinuous culture of algae 167
Horizontal water bottle 14, 47, 121
Hot-wire instruments 18
Humic acids 166
Husmann's filter glass 132, *133*
Hydra 162

Hydrachnellae (Hydracarina) 80, 85, *127*, 160, 163
culture 163
fixation and conservation 160
Hydrobacteriology ix
Hydrobiology ix
Hydrobotany ix
Hydrocampa nymphaeoides 2, *3*
Hydrochares 75, 77
Hydrogen-ion concentration 21, 139
see also pH value
Hydrometric measuring vanes 18
Hydroxylamine 160
Hydrozoology ix
Hygropsammon 71, 129
Hypersaprobic conditions 150
Hypnum schreberi 164
Hypolimnion 3, 70
Hyponeuston 67
Hyporheic diggings 118
Hyporheic distribution of organisms 128
Hyporheic share of life 8, 104, 109, 113, 118, 128, 130, 131

Idranel 30
Indicator organisms 148, 149, 151
Indicator values 154
Indicators for estimation of pH 21
Inhabitants of our smallest cavities 129
Inorganic nutrients 134
Insect larvae 83, 99, 159
Internal friction of water 4
Interstitial ground water 28, 70, 71, 75, 129
Interstitial inhabitants 120
Interstitial spaces 118
Iron rakes 80
Isopoda 130
Isosaprobic conditions 150

Jenkin surface and sampler 93

Keratella *5*, *56*, 63, 64, 164
Koenike mixture 160
Kolkwitz chamber *59*
Kolkwitz's chamber techniques 59
Kolkwitz's mud sieve 95

Lake type system 84
Lake water, stratification of 3
Lakes in caves 1, 129
Lampropedia halina *67*
Leaf miners 2, 80
Lepidoptera 160
Leptodora kindtii 145
Light 16, 76, 129, 139, 167
ecological significance of 2, 4, 70, 76
physiological significance of 137
Light–dark change in algae 167
Light shock in algae 129
Light-thermostats for algae 167
Limnohalacaridae 164
Limnology ix
Limnosaprobic conditions 150
Liponeura 8, 104, 111
Lobohalacarus weberi quadriporus 163
Long-handled scraper *74*, 75, 96, 112
Lugol's solution 139, 161
Lux (unit of light intensity) 17

Macroplankton 33
Manufacturing firms 168
Marine biology ix
Marking of drawlines 14
Measurement of radiation under water 16, 141
Mechanical stage 51
Megaloptera 160
Megaplankton 33
Melosira varians 83
Membrane filter 137, 140
Membrane filtration 64
Menthol crystals 160
Meschkat method 76
Mesh-width 38, 39, 95, 113, 117
Mesocyclops leuckart 72
Mesoplankton 33
Mesopsammon 129
Mesosaprobic zone 8, 149
Metabolic cycle in lakes *7*
Metabolism of water bodies 70, 134
Metalimnion 3, 70
see also Thermocline; Transition zone
Metal cylinder
for shore-diggings 131
for shore-grabs 131, *132*
Metal sieve for the analysis of sediment 95
Metasaprobic conditions 150
Method
of Douglas 107
of Goldman 141
of Karaman–Chappuis 118
of Knopp 151
of Liebmann 155
of Margalef 107
of Pantle and Buck 153
of Remane 72, 161
of Zelinka and Marvan 155
Meyer's sampler bottle *11*
Microaquaria 163
Microplankton 33
Microregma heterostoma 158
Microscope inverted 52, *58*, 64
Mineral nutrient solution 165

Mixodiaptomus laciniatus, culture of *66*, 162
Moina 162
Monaco grab *89*, 91
Monadodendron 83
Monodur bolting cloth 38
Monofilament bolting cloth 38
Mosses 104, 106, 108
Mougeotia 162
Mud sampler 74, 85, 88, 92
  of Apstein 92
  of Elgmork 92
  of Naumann 92
Mud sieve of Kolkwitz 95
Munich method 155
Myriophyllum 75, 163

Nais 160
Nannoplankton 6, 33, 54, 55, 75, 145, 161, 167
Narcotization 160
Nautococcus emersus 67
Navicula 67
Nekton 6, 8
Nematoceran larvae 99
Nematodes, preservation of 76, 80, 85, 99, 160, 162
Nemertina 160
Net dredges 77
Net factor 42
Net funnels 38, 40, 41
Net opening 41
Net plankton 33, 52, 55
Nets 37, 75, 118, 130
Neuroptera 160
Neuston *67*
Nitrogen method 137
Normal net 38, *40*, 55
Nostoc 80
Notodromas monacha 67
Nuphar 75
Nutrient media of algae 165
Nutrient solution No. 10 of Chu 165
Nylon gauze 38, 123
Nymphaca 75

Odonata 160
Open system 7, 104
Oligochaetes 85, 99, 160, 163
Oligosaprobic zone 8, 149
Oligotrophic lakes 63, 84
Opal glass method 167
Ophrydium 80, 83
Optical properties of water 4, 16, 70
  see also Light; Vertical migration of plankton
Organic pollution 157, 158
Organismal drift 121
Oribatidae 160
Oscillatoria 52, *56*, 150
Osmic acid 159
Osmium tetroxide 68
Oxygen bottle 26
Oxygen calculator *26*
Oxygen consumption
  at the bottom of waters 4, 84
  of bacteria 137
Oxygen method 137
Oxygen saturation concentration 23

Particle size analysis 20, 72, 120
Perforated metal collecting tubes *131*
  see also Dredges
Peridinium 52, *53*, *56*, 164
Periphyton *see Aufwuchs*
Perlon drawlines 41
Petersen 85, 118
Petit tube for measurement of the water current 18
Pfeiffer's mixture 106, 161
pH value 21, 139
Photoautotrophic plants 69, 84, 134, 137
Photometer 16, 141, 158
Photometric determination of the number of cells 158
Photometric measurements (in chemistry) 28
Photoperiodicity 85
Photosynthesis 6, 64, 83, 141, 142, 145, 164, 167
Phragmites communis 75, 76
Phreatic organisms 129
Phreatobiontic 130
Phytoplankton 34, 52, 59, 121
  investigation of 55
Piscicola 169
Pisidium 85
Piston-borer (in section on mud samplers; etc.) 93
Piston pipette 49
Planarians 130
Plankton 33, 75, 80, 121
Plankton clouds 35
Plankton horizon (layer) 35, *36*
Plankton nets 37, 41, 42, 64, 75, 79, 130
Plankton organisms, distribution of, in lakes 6, 34, 35
Plankton sampler of Clarke and Bumpus *45*, 93
Plankton subsampler 49
Plankton, vertical migrations of 4, 34, 38, 80
Plant communities, charting of 80
Plant hooks 106
Plant mat 111
Plant stems 76
Plant volume 111
Plant water 1
Plastic bands 125
Plastic films *124*, 125, 142
Plecoptera 99, 122, 160
Pollution of water 147
Polyethylene bottles 10, 107, 120
Polygonum amphibium 77
Polysaprobic zone 7, 149

Polytrichum (moss) 164
Pond-skater *(Velia)* 67
Pond skaters (Gerridae) 67
Porifera 160
Porous rocks 129, 131
Porous spaces 71, 129, 131
Potamogeton alpinus 77
Potamogeton natans *2*, *77*, 80
Potamogeton zone 70
Potamon 105
Potassium iodide–iodine solution (Lugol's solution) 58, 139, 161
Potassium permanganate consumption 31, 157
Predators 134
Preservation of zooplankton 48
Pressure (hydrostatic) 75
Primary consumers 7, 134, 145
Primary producers 6, 65, 135, 157, 145
Primary production 3, 6, 147, 157
Production biology ix, 134
Production capacity *142*
Production of littoral plants 70, 134
Profundal zone 6, *69*, 77, 84 ff., 113
Progressive choking of nets 55
Protein content of algae 164 ff.
Proteus anguineus 129
Protozoa 99, 161
Psammon 131
Pseudomonas fluorescens 158
Pump collections 80
Pump wells 131
Pure culture of algae 164
Putrefaction 4, 27, 147, 149, 158
  see also Decay
Pyramid traps 122, *123*

Quadrangular dredge 78
Quality of water, classes of 150

Radioactive carbon 6, 138 ff.
  see also C^{14}
Radioactive contamination 149
Radioactive-marked algae 145
Radioactive sodium bicarbonate 139
Radioactive waste water 150
Radioactivity 138
Radiosaprobic conditions 150
Ranunculus 106
Raphidium 164
Rate of emergence of insects in streams 123
Reducers 148, 157
Reed swamp 69, 70, 75 ff.
Regeneration coefficient 146
Regions of running water 104, 105
Reineck's box grab 85, 89
Relative contamination-load of water 153
Relative purity of water *152*, 153
Respiration 83, 141
Respiration intensity 1
Respiratory loss 141
Resting stage of Cyclopidae 85
Reversing thermometer 15
Rhitrogena 111
Rhitron 105
Rieth's plant grab (collector) 80
Rotary pumps 47
Rotifers (culture) 83, 130, 161, 163
RPC system of Gabriel 157
Ruttner's scissor grab 89

Salmonidae, region of 105
Sampling cylinder of Neil 113, *114*
Sand sample 72
Saprobial condition
  degree of 155
  index of 153
  stages of 153
Saprobial system 148
  of Fjerdingstad 149
  of Kolkwitz and Marrson 148
  revision of, by Liebmann 149
  of Sladecek 149
Saprobial valency 154
Sapropelic world of life 148
SBV (alkalinity) capacity for combination with acids 30, 139, 140
Scapholeberis 67, 72
Scapholeberis kingi 72
Scenedesmus quadricauda 158, 162, 167
Schaudinn's solution 160
Schlee's revolving collector 103
Scirpus lacustris 75
Secchi disc *17*
Secondary consumers 134, 145, 146
Sediment activity of Caspers 158
Sedimentation 55
  of plankton 60 f., 64
Sedimentary volume 63
Sediments, stratification of 89–92
Self-absorption of algae 142
Self-purification 147, 157
  autotrophic phase of 147, 158
  heterotrophic phase of 147, 158
  of running water 8
  oxidative phase of 147, 157
  photosynthetic phase of 147, 158
Sergentia coracina 85, 102
Shore area 69
Shore bank *69*, 70
Shore, diggings on 131
Shore flight of plankton 34
Short-circuited metabolic cycles 134

Shovel-sampler 113, 115, *116*
Sialis 85
Sieve, sets of, for the analysis of sediments 19, 95
Sieve-box
  of Jonasson 95
  of Kolkwitz 96, 97
Sieve bucket of Rawson and Fleming 95
Sieving of bottom samples 95 ff.
Silicate content of water 10
Silk gauze 33
Simuliidae (black flies) 8, 123
Simulium 111, 123
Sinking, rate of 4
  see also Flotation techniques; Form resistance; Plankton
Slope down from the shore 49, *69*
Snails 160
Soil extract 166
Sorting apparatus of Moon *125*
Species deficit of Kothe 148, 156
Specific denser solutions for collecting from bottom samples 99
Speed of flow 18, 104, 156
  see also Current; Water, movement of
Sphaeriidae 160
Sphaerotilus 106, 150
Spherical curves 65
Spirogyra 70, 80
Spongillidae 160, 164
Spring (well) water 22, 27
Square foot stream bottom sampler *114*
Stagnation periods 84
Stalactites 129
Stalagmites 129
Standing crop 135
Stationary phase of algae 167
Staurastrum *5, 53, 56*
Stenus 67
Sterile by filtration 165
Sterilization in the steam-chamber 165
Stigeoclonium 106
Stinging nettle powder 163
Structural type (in underground water organisms) 8
Stygobionts 129
Stylaria 160
Sublittoral zone 70
Submerged meadows 70, 75, 80
Subterranean sphere of life 8–9
Suckers 8, 104
Suction filtration 140
Supplementary consumption of Knopp 158
Surf on the shore 70, 72, 75
Surface of stone (rocks) 74, 75, 111
Surface-area methods 49
Surface film of water 6, 48, 67
Synchronous culture 167
Synedra 83
Synura 166

Tabellaria *5*, 52
Tables of Harvey and Rodhe 140
Tanytarsus 84, 95
Tanytarsus lakes 85
Tardigrades 161
Temperature
  amplitude of 105
  physiological action of 146, 156, 165
  tolerance of 105
  underground water 129
  water 14, 104
Tendipes (= *Chironomus*) 84, 95, 102
Tesa film 92
Testacea 76, 80
Tetraspora cylindrica 83
Thermocline 3, 48, 70
  see also Metalimnion
Thermoelements 15
Thiamin 166
Thienemann net *110*, 111
Titisee (Black Forest) 85
Titriplex 31
Titrisol ampoules 120
Total carbon 139
Total hardness of water 31
Touch sense in cave animals 129
Toxicity of propellants (motor fuels) and oils 158
Trace-element solutions 166
Transition zone 3, 70
Trans-saprobic conditions 150
Travertin 22
Tree-cavities 1
Trichoptera (conservation and fixations) 99, 111, 122, 161
Tricladidae 161, 164
Troglobiont 129
Trophic grade of water-bodies 148, 158
Trophic pyramid 134, 135
Trophogenic zone 41, 70, 84, 134, 145
Tropholytic zone 4, 70, 84, 134
Trout 105
Toxic waste water 149
Toxic zone 149
Tube-chamber 59, *60*, 61, 64
Tube sampler
  Garnet and Hunt's 80
  Liebmann's 92
  Zullig's 93
Tubificidae 95, 164
Turbellaria 160
Typha 75, 76

UHU-plus 120
Underwater television camera *35*
Underwater emergence-traps 102
Ulothrix 106
Ultra-nannoplankton 33
Ultraplankton 33, 54

Ultrasaprobic conditions 150
Unicellular plankton algae 164
Unit of population of stones (rocks) 112

Vaucheria 106
Velia 67
Vertical capture 52
Vertical distribution of bottom fauna 88
Vertical migration of plankton 4, 34, 38, 80, 85
Vertical population density 65
Vertical profile 42, 48, 65, 88
Viscosity of water 5
Visibility, depth of, in water 17
Volume of plankters 63
Volvox 52, *53*
Vrana lake 70

Warm-tone lamps 167
Waste water (sewage) 147
Water
  movement of 104, 121, 123
  purity or quality of 147
  thrust of 104
Water-bottles, Ruttner's *12, 13*
Waterglass 99
Water knot grass *(Polygonum amphibium)* 77
Water mites 80, 85, 113, *127*, 135
  culture of 163
  fixation and preservation 160
Water pumps 38, 47, 55, 120
Water-soluble phosphate 29
Water strider *(Hydrometra)* 6, 67
Winter stagnation 3
Wire sieve 95
Worms 83, 160, 163

Xenosaprobic zone 154

Yeast 162
Yeast extract 166

Zeppelin net 41, 55, *57*
Zonation of running water 126
Zooplankton 4, 34, 55, 145, 156
  see also Plankton